Effects of the Blended Retirement System on United States Army Reserve Participation and Cost

Beth J. Asch, Michael G. Mattock, James Hosek

Prepared for the United States Army

Approved for public release; distribution unlimited

For more information on this publication, visit www.rand.org/t/RR2591

Library of Congress Cataloging-in-Publication Data is available for this publication.
ISBN: 978-1-9774-0238-7

Published by the RAND Corporation, Santa Monica, Calif.

Cover: U.S. Army Photo.

Preface

This report documents research and analysis conducted as part of a project entitled *Army Reserve Transition to the Blended Retirement System*, sponsored by Office of the Chief, Army Reserve. The purpose of this study was to provide estimates of the change in costs for the U.S. Army Reserve due to the Blended Retirement System in the transition years and the number of participants as a result of both new entry to the Army Reserve and election ("opt-in") by serving members.

The Project Unique Identification Code for the project that produced this document is HQD167573.

This research was conducted within RAND Arroyo Center's Personnel, Training, and Health Program. RAND Arroyo Center, part of the RAND Corporation, is a federally funded research and development center sponsored by the United States Army.

RAND operates under a "Federal-Wide Assurance" (FWA00003425) and complies with the *Code of Federal Regulations [CFR] for the Protection of Human Subjects Under United States Law* (45 CFR 46), also known as "the Common Rule," as well as with the implementation guidance set forth in U.S. Department of Defense (DoD) Instruction 3216.02. As applicable, this compliance includes reviews and approvals by RAND's Institutional Review Board (the Human Subjects Protection Committee) and by the U.S. Army. The views of sources utilized in this study are solely their own and do not represent the official policy or position of DoD or the U.S. government.

Contents

Figures

Tables

Summary

The Blended Retirement System (BRS), created by the National Defense Authorization Act (NDAA) of 2016, represents the first major change to the armed services' retirement system since the end of World War II. The BRS retains a defined-benefit (DB) plan from the legacy system and adds a defined-contribution (DC) plan known as the Thrift Savings Plan (TSP) that vests much earlier than the DB plan and provides contribution matching up to 5 percent of basic pay, and a continuation pay (CP) at midcareer as a retention incentive. The CP a member receives is determined by a service-set CP multiplier times the member's monthly full-time basic pay. Members reaching retirement under the BRS have the option to receive part of their DB annuity as a lump-sum payment payable immediately upon retirement from the military. Members who were serving as of December 31, 2017, are grandfathered under the legacy system, but members with fewer than 12 completed years of service (YOS) (or reservists with fewer than 4,320 retirement points) at the start of BRS implementation on January 1, 2018, have the opportunity to opt in to the BRS. The reserve retirement benefit is based on creditable years of service, which equals the number of retirement points a reservist accumulates divided by 360. Reservists accumulate points for participating in reserve drills, inactive duty training, active duty, and other activities including funeral honor guard service. Participating reservists also earn 15 participation (or "gratuitous") points per year. The active retirement benefit is based on total YOS, where a year equals 365 points. Thus, the BRS can have impact on the retention and participation behavior of current members of the active component (AC) and reserve component (RC), as well as on members who join after the start of BRS implementation and are automatically enrolled in the new retirement system.

The move to a new retirement system creates uncertainty for the Army about its ability to sustain Regular Army (RA), U.S. Army Reserve (USAR), and Army National Guard (ARNG) strength, how the BRS might affect flows of RA members to the RC, whether the experience mix of the Army would change in a way that no longer meets Army readiness requirements, and how the transition to the BRS would affect personnel costs in the short and long run. The transition to the BRS also raises questions about how the CP should be set by the Army for the RA, USAR, and ARNG, and how alternative CP policies would affect opt-in behavior, RA retention, USAR and ARNG participation, and cost.

The Office of the Chief of the Army Reserve (OCAR) requested that we address the question of the impact of the BRS on the USAR under alternative courses of action (COAs) regarding setting the CP multiplier; while the BRS legislation sets a minimum CP multiplier for the RC of 0.5, the Secretary of the Army has discretion to raise the multiplier above this minimum. The OCAR wanted to know how the BRS under alternative CP multiplier COAs would affect USAR participation, AC retention, the flow of members from the RA to the USAR, personnel costs, and opt-in behavior among currently serving USAR members. Thus, the focus of our analysis is on the USAR, though we also provide selected results for the RA and ARNG because our modeling accounts for flows between the three Army components.

Past RAND Corporation analysis provided estimates of the effects of the BRS on AC retention and cost, as well as RC participation for officer and enlisted members of all armed services (Asch, Hosek, and Mattock, 2014; Asch, Mattock, and Hosek, 2015). The analysis was conducted using the dynamic retention model (DRM) and modules written for simulation, graphics, and costing. The DRM is a model of an individual's retention decisions over the individual's active and reserve careers. The DRM accounts for expected military and external earnings, allows for individual differences in one's taste for military service and for random shocks in each period, and incorporates the ability of individuals to reoptimize depending on the conditions realized in a period. Models were estimated for each service, with separate models for officers and enlisted personnel, using longitudinal retention data. The estimated models were used to simulate the retention and cost effects of changes to the compensation system, including the BRS in the steady state and the transition to the steady state. By *steady state*, we mean when all members have spent an entire career under one system, for example, the new reformed system.

In our earlier analyses, the DRM did not distinguish between National Guard and reserve service for the Army and Air Force (the two services with both Guard and reserve components) in assessing the effects of the BRS on the RC and analysis of the transition effects of the BRS. In particular, estimates of the percentage of personnel that would opt in and the time path of costs were for the AC only, not the RC.

Our approach in this report involved extending the DRM capability used in previous studies to distinguish between USAR and ARNG service so that we could explicitly assess how the BRS would affect the USAR. This involved extending the mathematical structure of the DRM and estimating new Army models for enlisted personnel and officers. We estimated these models using individual-level data on RA, USAR, and ARNG members who entered the Army as non–prior-service entrants in 1990 or 1991 and tracked their service through 2015, providing up to 26 years of data on each individual. The data were provided by the Defense Manpower Data Center. We found overall good model fits for our extended models. We also extended the simulation capability to enable us to assess how the BRS would affect the USAR in the transition as well as the steady state. We considered four COAs for setting the CP multiplier. Specifically, we considered setting the multipliers:

1. at the mandated floors of 2.5 for AC and 0.5 for RC
2. at levels for the RA, USAR, and ARNG that sustain retention in all three components relative to the baseline, where the baseline is assumed to be retention under the legacy retirement system

3. at levels for the RA and USAR that sustain retention in these two components relative to the baseline, assuming the ARNG acts independently and sets its multiplier to the mandated floor
4. at levels for the USAR that sustain retention in that component relative to the baseline, assuming the RA and ARNG act independently and set their multipliers to the mandated floors.

While there is no presumption that future requirements will call for the same size and mix as the baseline force under the legacy retirement system, we assessed the retention effects of the BRS and considered the CP multipliers in COAs 1–4 in terms of how well they could achieve the baseline. To identify the multipliers that would sustain retention in COAs 2–4, we used the DRM simulation capability to develop a computer optimization routine that finds the CP multipliers that minimize the distance between the retention profiles in the baseline and the profiles under the COA (e.g., that most closely replicate the baseline retention profiles in each component). The optimized CP multipliers under each COA are shown in Table S.1, as is the percentage change in force size for each component. The multipliers in the table assume the CP is paid at 12 YOS for a four-year obligation. NDAA 2017 changed CP to allow the services to pay it at any YOS between eight and 12 and reduced the obligation to three years. Our 2017 study of the BRS for the Army and the other services assessed the BRS under NDAA 2016, and did not consider the NDAA 2017 changes. To be consistent with our 2017 analysis of the BRS for the Army, our current analysis for the USAR assumes that the CP multiplier is paid at YOS 12 and requires a four-year obligation. Additional analysis would be required to assess how changes in the targeted YOS would change the optimized CP multipliers and CP costs. In the case of the RA, force size is held constant by increasing or reducing accessions as retention changes under each COA. Thus, the table shows no change in RA force size, by assumption.

Our results show that under COA 1, the BRS can support a steady-state force for the RA, USAR, and ARNG that is quite close to the current forces for enlisted personnel, but not for officers in each component. Because the mandated floors for the CP multipliers sus-

Table S.1
Continuation Pay Courses of Action, Optimized CP Multipliers, Steady-State CP Costs, Percentage Change in Force Size

COA	Multipliers			CP Cost (2017 $M)				Percentage Change in Force Size Relative to Baseline		
	RA	USAR	ARNG	RA	USAR	ARNG	Total	RA	USAR	ARNG
Enlisted[a]										
1. RA, USAR, ARNG set to floor	2.50	0.50	0.50	64.7	7.3	17.7	89.7	0%	0.1%	0.6%
Officer										
1. RA, USAR, ARNG set to floor	2.50	0.50	0.50	40.2	2.2	1.1	43.5	0%	5.5%	7.2%
2. Optimize RA, USAR, ARNG	8.28	2.25	1.08	138.0	9.3	2.4	149.7	0%	–0.2%	–3.0%
3. Optimize RA, USAR; ARNG set to floor	7.59	1.78	0.50	126.1	7.4	1.1	134.6	0%	0.1%	–3.1%
4. Optimize USAR; RA and ARNG set to floor	2.50	0.19	0.50	NA[b]	NA[b]	NA[b]	NA[b]	NA[b]	NA[b]	NA[b]

NOTE: NA = not applicable.
[a]COAs 2–3 are not relevant for enlisted personnel, because retention is sustained under COA 1.
[b]COA 4 is infeasible so CP costs and force size changes were not estimated.

tain retention, we do not assess COAs 2–4 for enlisted personnel. For officers, we found that the mandated floors are too low; officer RA retention is too low relative to the baseline, and USAR and ARNG participation levels are too high. That is, fewer officers complete an RA career and more participate in the USAR and ARNG. For officers, higher CP multipliers—close to about eight months of basic pay for RA personnel and about 12 months of basic pay for USAR personnel—are required. Specifically, under COA 2, the simulations indicate that CP multipliers should be 8.28, 2.25, and 1.08 for RA, USAR, and ARNG officers, respectively, to jointly sustain force size in all three components. Under COA 3, they should be 7.59 and 1.78 for RA and USAR officers, respectively (and assumed to be 0.5 for ARNG).

We find that under COA 4, the CP multiplier that sustains USAR retention is 0.19, a figure that is below the mandated floor of 0.5. Thus, this COA is infeasible. If the multiplier is, instead, set to this floor, COA 4 and COA 1 are identical. Since we already assess COA 1, we do not provide additional results for COA 4.

In addition to retention effects, the DRM also provides estimates of CP costs under COAs 1–3. Because the Department of Defense (DoD) actuary provides estimates of the change in terms of DB and TSP costs, the focus of our cost analysis was on CP costs. OCAR asked us to provide estimates of CP costs under each COA in the steady state and in the transition years.

Table S.1 also shows estimated steady-state CP costs in millions of 2017 dollars for each component. When CP multipliers are set at the floors under COA 1, CP costs are $89.7 million for enlisted personnel overall and $7.3 million for enlisted personnel in the USAR. For officers, CP costs are $43.5 million overall in the steady state and $2.2 million for the USAR. But under COA 1, RA officer retention is not maintained, so RA officer accessions must increase to sustain the force, and USAR and ARNG participation under COA 1 exceeds participation under the baseline compensation system. Sustaining retention across the three components means that officer CP costs are about three times higher overall in the steady state, $149.7 million under COA 2 (and $134.6 million under COA 3). For the USAR, officer CP costs are more than four times higher under COA 2 than COA 1, $9.3 million versus $2.2 million.

In addition to the CP costs, each armed service will also be required to make TSP contributions on behalf of members, another source of steady-state cost. Offsetting these costs are the savings to the government associated with lower DB payouts and therefore lower DB accrual charge in future years.

In the transition to the BRS, we estimate that 56 percent of USAR enlisted personnel with prior AC service will opt in to the BRS, compared with 37 percent of enlisted personnel without prior AC service. The percentage is lower for the latter group because the group has a relatively low chance of benefiting from the elements of the BRS. This group also has a low chance of benefiting from the legacy system, given its lower retention, so at the margin, there is no incentive to choose to elect the BRS. For example, about half of entrants without prior AC service would reach the start of YOS 3, when TSP matching contributions begin and when TSP vesting occurs, and fewer than 10 percent of entrants would reach YOS 12 when they would receive CP.

For USAR enlisted personnel with prior active service, predicted opt-in behavior varies with years of service at the time of the opt-in decision, as follows:

- Virtually every junior member with five or fewer YOS would choose the BRS over the legacy system. This is because junior members have a low probability of vesting under the legacy system, so the BRS with earlier vesting in the TSP is more attractive.
- Beyond YOS 5, the percentage of personnel at each YOS that would opt in decreases with YOS. These more senior personnel have missed out on the Army's TSP contributions on their behalf had they entered the BRS when they were more junior. Consequently, all else being equal, the BRS is less attractive to those with more YOS.

Because it is possible that some USAR personnel have more than 12 YOS but fewer than 4,320 retirement points, we estimate that some USAR enlisted personnel with more than 12 YOS elect the BRS, though still relatively few compared to those with five or fewer YOS.

For USAR officers, predicted opt-in rates vary with the CP course of action. When CP multipliers are set high enough to sustain retention under COAs 2 and 3, the opt-in rate is predicted to be relatively high, 30 percent and 25.6 percent, respectively. But we found that few officers (just 15.4 percent) would opt in to the BRS when CP multipliers for officers are set at minimum levels under COA 1. Like enlisted USAR personnel with prior AC service, USAR officer opt-in rates are highest among those with the fewest YOS who have the most years to benefit from TSP contributions made on their behalf and from CP.

DoD released preliminary opt-in rates for USAR and ARNG personnel in January 2019, indicating that USAR opt-in was about 12 percent and ARNG was about 10 percent, significantly below the rates estimated with the DRM. The reason for the lower than predicted opt-in rate is unclear. The DRM assumes members fully understand the elements of the BRS and the financial implications of choosing it over the legacy system and that they choose the system yielding the highest expected payoff, accounting for lack of perfect information about the future. It is possible members did not fully realize the importance of the opt-in decision or understand the features of the BRS, or they relied on input from influencers who may be older and closer to retirement and who provided input that was not in the best interest of individual members. Another possibility is that individual members, although informed about the BRS and the opportunity to opt in, were not required by the service to make a choice. The services differed in their approaches regarding the opt-in decision, and only the Marine Corps required marines to make a choice either to opt in or not. Requiring a choice might have made the decision more salient and induced service members to focus attention on it. The differences between the predicted and (preliminary) actual opt-in rates suggest that the DRM, despite fitting the data well and predicting retention behavior well, may require additional data and information for projecting choices about a new compensation system. More information is needed to better understand the factors and people that influenced members' choices.

We also estimate how USAR CP costs change over time during the transition years of the BRS. We find that the time pattern of CP costs is not smooth but that costs jump up after 12 years, when new entrants in 2018 reach eligibility for CP. Growth is quite slow in the first 12 years under COA 1, and the increment after twelve years is larger, because relatively few USAR members who are grandfathered in under the legacy system elect the BRS.

These results imply that if short-term cost considerations are of primary importance, the Army would set the multiplier at the floor for each component and address retention later when retention issues emerge. The upside of this strategy is that the CP costs increase more slowly. The downside is that opt-in rates are lower, so the future cost savings of lower DB costs will also be realized more slowly. Alternatively, if longer-term cost savings are of primary importance, the results imply the Army would set the multipliers for officers at the higher levels required to sustain retention. At these higher levels, opt-in rates increase, and the cost savings of a lower DB accrual charge and DB outlays are realized more quickly.

Finally, the DRM capability we developed for this study considers the interactions between the Army components and the effects of compensation changes in one component—such as the RA—on the other components, notably the USAR. This capability could be used in the future to assess the effects of other compensation policies of interest, such as assisting the OCAR in ensuring that special and incentive pays and bonuses are in place to sustain USAR force size and justifying changes to legislated caps on these pays.

Acknowledgments

We would like to thank our Army sponsor, Robert M. "Mike" Maxwell, chief financial officer and director of Resource Management and Materiel for the Office of the Chief of Army Reserve. We are grateful to our action officer, Gary Morris, deputy director, Army Reserve Program and Analysis and Evaluation Division at the Office of the Chief of Army Reserve. Gary Morris provided excellent input throughout the course of the project. We also benefited from the input of LTC Marken Orser and LTC Micah "Buddy" Bright, both manpower analysts within the Program Analysis and Evaluation Division in the Office of the Chief of Army Reserve. At RAND, we wish to thank Arthur Bullock for outstanding data programming. We also thank Bryan Hallmark, Shanthi Nataraj, Michael Hansen, and Michael Linick within RAND Arroyo Center who provided guidance during our project. We also appreciate the excellent programming assistance we received from Anthony Lawrence. Our report also benefited from the input and comments of two reviewers: Curtis Gilroy, formerly director of the 9th Quadrennial Review of Military Compensation and director of the Office of Accession Policy within the Office of the Secretary of Defense; and our RAND colleague Matthew Baird. We greatly appreciate their help.

Abbreviations

AC	active component
ARNG	Army National Guard
BAH	basic allowance for housing
BAS	basic allowance for subsistence
BRS	Blended Retirement System
COA	course of action
CP	continuation pay
DB	defined benefit
DC	defined contribution
DMDC	Defense Manpower Data Center
DoD	U.S. Department of Defense
DRM	Dynamic Retention Model
FY	fiscal year
NDAA	National Defense Authorization Act
NPS	non-prior service
OCAR	Office of the Chief of the Army Reserve
PS	prior service

RA	Regular Army
RC	reserve component
RMC	regular military compensation
TSP	Thrift Savings Plan
USAR	United States Army Reserve
WEX	Work Experience File
YOS	years of service

CHAPTER ONE

Introduction

For decades, the armed service has operated under a defined-benefit (DB) system that provides an annuity set at 2.5 percent multiplied by years of service (YOS) multiplied by basic pay, with vesting occurring after 20 YOS. The military retirement system differs for the active component (AC) and reserve component (RC). Retirement benefits start immediately upon retirement for AC personnel, but not until age 60 for reservists (or somewhat sooner depending on the extent of a reservist's deployment).[1] Further, the reserve retirement benefit is based on creditable years of service, which equals the number of retirement points a reservist accumulates divided by 360. Reservists accumulate points for participating in reserve drills, inactive duty training, active duty, and other activities including funeral honor guard service. Participating reservists also earn 15 participation (or "gratuitous") points per year. The active retirement benefit is based on total years of service, where a year equals 365 points.

The National Defense Authorization Act (NDAA) for fiscal year (FY) 2016 created a new retirement system for the armed services called the Blended Retirement System (BRS) that continues to include a DB plan for the AC and RC but adds two new components, a defined-contribution plan (DC), known as the Thrift Savings Plan (TSP), and continuation pay (CP). The TSP provides an automatic agency contribution on behalf of service members with additional matching contributions. TSP vests after two YOS, much earlier than

[1] NDAA 2008 enacted legislation that allows RC members to retire as early as age 57 for time deployed in support of a national emergency.

the 20 years required to vest in any retirement benefit under the legacy system. Continuation pay is a retention incentive paid to members in midcareer who commit to a service obligation. As a tradeoff to adding the TSP and CP components, the NDAA reduced the DB multiplier from 2.5 percent to 2.0 percent. A key role of CP is to provide a retention incentive among those in their midcareer to offset the reduction in retention incentives that would accompany the reduced DB multiplier. Members who qualify for the DB have the option to receive part of the DB annuity they receive between their retirement age and age 67 (or the Social Security retirement age) in the form of a lump sum; this will be described in more detail in the next chapter. All new accessions after January 1, 2018, are automatically enrolled into the BRS. Members serving as of December 31, 2017, were grandfathered into the legacy system while those with fewer than 12 YOS in the AC or 4,320 retirement points in the RC have the opportunity to opt in to the BRS in 2018.[2]

The move to a new retirement system creates uncertainty for the Army in several ways: in the system's ability to sustain Regular Army (RA), United States Army Reserve (USAR), and Army National Guard (ARNG) strength at the same levels; in how the BRS might affect flows of RA members to the RC; in knowing whether the experience mix of the Army would change in ways that no longer meet Army readiness requirements; and in how the transition to the BRS would affect personnel costs in the short and long run. The transition to the BRS also raises questions about how the CP should be set by the Army for the RA, USAR, and ARNG, and how alternative CP policies would affect opt-in behavior, RA retention, USAR and ARNG participation, and cost.

---

2 The figure 4,320 was chosen because 4,320/360 is 12 years.

Office of the Chief of the Army Reserve Asked RAND Arroyo Center to Assess the Effects of the Blended Retirement System

The focus of this report is on the USAR. The Office of the Chief of the Army Reserve (OCAR) requested that we address the question of the impact of the BRS on the USAR under alternative courses of action (COAs) regarding setting the CP multiplier. The OCAR wanted to know how the BRS under alternative CP multiplier COAs would affect USAR participation, AC retention, the flow of members from the RA to the USAR, personnel costs, and opt-in behavior among currently serving USAR members. OCAR questions included whether the new policy would enable the USAR to sustain its current force size; whether there would be changes in experience mix; whether the flow from the RA to the USAR would change; and whether there would be repercussions on the RA force.

Intuitively, the lower DB multiplier under the BRS would reduce USAR participation before 20 YOS, while the addition of an early-vested DC plan and CP would improve participation. Whether these effects are offsetting is an additional question of interest. The BRS could also change the USAR experience mix, depending on how CP multipliers are set. Midcareer RA retention might decrease if the RA CP multiplier is set too low, and individuals who would have stayed in the AC might flow to the USAR, increasing USAR midcareer participation under the BRS. Alternatively, if CP in the AC is set too high, more officers and soldiers will want to stay in the RA, and USAR participation in midcareer might decline. Another question is whether the BRS would cost more or less to the Army than the legacy retirement system and how cost is affected by CP policy in both the short and long term.

The purpose of this study is to address these questions, making use of and further developing a model the RAND Corporation created of AC and RC retention, known as the dynamic retention model (DRM), that permits simulations of new and untried policies. The focus of the analysis is on how the BRS would affect the size and shape of USAR forces, as well as cost, relative to the legacy retirement system.

Because our approach, outlined next, incorporates flows between the RA, USAR, and ARNG, we also provide selected results for the RA and ARNG.

Overview of Approach

RAND's DRM is well-suited to the analysis of structural changes in military compensation such as the BRS. Earlier applications of the model included analyses in support of the 10th and 11th Quadrennial Review of Military Compensation (Asch et al., 2008; Mattock, Hosek, and Asch, 2012) as well as for a Department of Defense (DoD) working group on military compensation reform, convened between September 2011 and June 2012, that recommended that the current military retirement system be modernized with a blended system (Asch, Hosek, and Mattock, 2014). The alternatives the working group considered were forwarded to the Military Compensation and Retirement Modernization Commission, an independent commission mandated by the NDAA for FY 2013, and the DRM was used to support the commission's deliberations (Asch, Mattock, and Hosek, 2015). The commission also recommended a blended system, and indeed, many features included in the legislated BRS came from the commission's recommendations. Finally, the DRM was used to assess the BRS on AC retention, RC participation, and cost for the Army, Air Force, Navy, Marine Corps, and Coast Guard (Asch, Mattock, and Hosek, 2017).

In all of these previous implementations of the DRM, the model did not distinguish between National Guard and reserve service when estimating the effects of compensation changes and retirement reform on the RC. For the Army, this meant that USAR and ARNG were combined into a single group, Army RC. Furthermore, our previous implementations provided estimates of retention and cost effects in the steady state for the AC and RC, and transition effects for the AC. That is, our DRM capability did not provide estimates of effects on RC participation and cost in the transition years or provide estimates of opt-in behavior for the RC.[3]

Our study addresses these two limitations of our earlier analysis for the purpose of analyzing the effects of the BRS on the USAR. Specifically, we estimated new Army models that distinguish between USAR and ARNG service in the DRM for Army officers and enlisted personnel. In addition, we extended the DRM capability to analyze the transitional effects of the BRS for the AC to also analyze these effects for the USAR. The expanded capability enabled us to provide estimates of the effects of the BRS on RA retention, on USAR participation among those with and without prior RA service, and on CP costs for both officers and enlisted personnel, in the steady state and the transition to the steady state.

We provide such estimates for four alternative courses of action for setting the CP multipliers. These courses of action include setting the CP multipliers at the minimum levels permitted by Congress and three versions of CP multipliers that are chosen so that in conjunction with the other elements of the BRS sustain USAR participation and RA retention in the steady state.

Organization of the Research Report

Chapter Two describes the features of the BRS. Chapter Three gives an overview of the DRM, describes the extensions of the estimation and simulation capability, and presents how well our new models fit the observed data used to estimate the models, with more model details provided in the appendix. Chapter Four discusses the courses of action we consider and presents steady-state simulation results, while Chapter Five describes results for the transition period. Chapter Six provides concluding thoughts. The appendix gives details on the data development and model adaptation for the USAR and presents estimates of the model's parameters.

---

[3] A notable exception is Asch, Mattock, and Hosek, 2015, where we show the effects of reserve retirement reform on Army RC participation over time.

CHAPTER TWO

Elements of the Blended Retirement System

The BRS has three main components: a DC plan, a DB plan, and CP. We describe each element and briefly discuss the opt-in feature of the BRS. The main elements of the BRS are summarized in Table 2.1, drawn from DoD (2017).

Revised Defined Benefit Plan

The revised DB plan has an annuity multiplier of 2.0 percent, down from 2.5 percent under the legacy system. That is, it changes the value of the retirement annuity from 2.5 percent × YOS × average of the highest three years of basic pay, to 2.0 percent × YOS × average of the highest three years of basic pay. Vesting for the DB plan continues to be upon completion of YOS 20. The legacy system is called the "high-three" system.[1] The lower portion of Table 2.2 shows examples

[1] Technically, there are actually three legacy systems in place as a result of modifications to the system in 1981 and 1988. Pre-1981 entrants receive a fully inflation-protected annuity that is computed based on final basic pay. Those who entered between 1981 and 1986 are under "high-three" where the retirement annuity is fully inflation-protected but based on the individual's high three years of basic pay rather than final basic pay. The Military Retirement Reform Act of 1986, known as REDUX, changed the annuity formula to (0.40 + 0.035 × YOS-20) × high-three average pay for the years between separation and age 62, at which time retired pay reverted to the high-three formula. REDUX also changed the inflation protection. As part of NDAA 2000, or TRIAD, members at YOS 15 who were covered by REDUX were given a choice to stay under REDUX and receive a $30,000 bonus or be covered by high-three. For simplicity, we assume the high-three retirement system in our analysis.

Table 2.1
Comparison of the Legacy and Blended Retirement Systems

Plan Element	Legacy	BRS
Defined-benefit vesting	20 YOS	20 YOS
Defined-benefit multiplier	2.5%	2.0%
Defined-benefit payment working years	Full annuity AC; NA RC	Full annuity or lump-sum option (50% or 25%); RC lump-sum based on annuity from age 60 to retirement age
Defined-benefit retirement age	NA AC; 60 RC	NA AC; 60 RC
Defined-contribution agency contribution rate		1% automatic; plus up to 4% matching (max = 5%)
Defined-contribution contribution rate YOS		1%: entry + 60 days until 26 YOS Matching: start of 3 YOS–26 YOS
Defined-contribution member contribution rate		3% automatic; full match requires 5% contribution
Defined-contribution vesting		Start of YOS 3
Continuation pay multiplier (months of basic pay)		Minimum 2.5 for AC, 0.5 RC; with additional amount varying
Continuation pay YOS/additional obligation		At 8–12 YOS with 3-year additional obligation
Opt-in		Must be serving on 1/1/2018 and have less than 12 YOS, or 4,320 points, as of 12/31/17; opt-in period is 1/1/18–12/31/2018

NOTE: NA = not applicable.

Table 2.2
Example Computations of TSP Contribution, Continuation Pay, and Retirement Annuity, in 2018 Dollars

	Active		Reserve	
YOS 12	**E6**	**O4**	**E6**[a]	**O4**[a]
Monthly basic pay	3,564	7,053	3,564	7,053
Annual TSP, individual contribution = 5%	2,138	4,232	340	672
Annual TSP, Army contribution = 5%	2,138	4,232	340	672
Annual TSP, total contribution = 10%	4,277	8,464	680	1,345
CP, if multiplier = 2.5	8,910	17,633		
CP, if multiplier = 0.5			1,782	3,527
YOS 20	**E7**	**O5**	**E7**[b]	**O5**[b]
Monthly basic pay	4,625	8,771	4,625	8,771
Annual legacy DB annuity	27,750	52,626	16,766	31,795
Annual BRS DB annuity	22,200	42,101	13,413	25,436

SOURCE: Authors' computations.
[a] Computation of TSP contributions for reservists assumes the reservist receives 62 days of basic pay, equal to 12 days for summer training and 48 weekend drills. Thus, annual pay for the purpose of computing RC TSP contributions is annual pay of an active member times 62/365.
[b] Computation of the reserve retirement annuity assumes the RC member has ten years of active service and ten years of reserve service where the reservists earn 75 retirement points per reserve year. Thus, for the purpose of computing the retirement annuity, YOS is computed as 10 x (1 + 75/360). Note that the reserve retirement formula assumes a full year is 360 points.

of the retirement annuity computation for AC and RC personnel retiring with 20 YOS. (The upper portion is discussed later in this chapter.) We assume the retiring enlisted member is an E7 and the retiring officer is an O5, and we further assume that the reservists served ten years in the AC and ten years in the RC, where each year in the RC earned the reservists 75 retirement points. Under these assumptions, YOS for the purpose of computing retired pay for reservists equals 10 × [10 × (75 ÷ 360)]. That is, years in the reserves are prorated by 75 ÷ 360. Given these assumptions, an enlisted AC member would earn $27,750 per year under the legacy DB system and $22,200 under the DB for the BRS. For officers, they would earn $52,626 and $42,101 for the legacy system and the BRS, respectively. The amounts are lower for the reservists, reflecting the prorated YOS in the annuity computations.

Upon AC retirement, members will be offered the option to receive the regular (2-percent multiplier) full annuity immediately or one of two lump-sum payment options—the member may choose either 25 percent or 50 percent of the discounted present value of future retirement benefits up to age 67—along with an offsetting reduced annuity up to age 67 and the regular full annuity thereafter. That is, all individuals would receive an annuity based on the 2-percent multiplier after 67, but for the period between the age of retirement and 67, the individual can choose at retirement to take a full annuity (no lump sum) or a reduced annuity with a lump-sum payment. The discount rate for computing the lump sum will be calculated by the Office of the Secretary of Defense each year and will be determined by adding 4.28 percent to the inflation-adjusted seven-year average of the Department of Treasury High-Quality Market Corporate Bond Yield Curve at a 23-year maturity.

Reserve component members can also elect a 25 or 50 percent lump sum at retirement, but retirement begins at age 60, unless deployment experience allows RC members to retire as early as age 57. The lump sum is a based on the discounted present value of future retirement benefits up to age 67, and at age 67, the individual receives the full annuity again based on the 2-percent multiplier.

We do not model the lump-sum choice. Instead, our analysis assumes that all members choose the full annuity and do not choose either of the lump-sum options.

Defined-Contribution Plan

The DC plan is known as the Thrift Savings Plan, or TSP. Under the BRS, all members joining the armed services after January 1, 2018, are automatically enrolled in the TSP with an automatic member contribution of 3 percent of basic pay. The service will contribute on behalf of the service member, and the service contributions consist of two parts: an automatic contribution of 1 percent of basic pay that begins 60 days after the start of service, and a matching contribution beginning at the start of the third year of service. The service will match up to 4 percent of basic pay, according to the schedule shown in Table 2.3. Members must contribute 5 percent of basic pay to receive the maximum 4-percent match. Service members opting in to the BRS in 2018 receive both the automatic and matching contributions immediately. Like other DC plans, members have full access to their TSP funds at age 59 and a half. Finally, members can contribute more than 5 percent of basic pay, but doing so does not result in a higher match rate.

Table 2.2 shows examples of the calculation of the annual TSP contribution for AC versus RC members using the 2018 basic pay table. The examples consider an E6 and an O4 with 12 YOS. Assuming an individual contributes 5 percent of basic pay, thereby receiving a 5-percent match from the Army, the total annual contribution to the TSP for the member would be $4,277 and $8,464 for active members who are E6 and O4, respectively, and would be $680 and $1,345 for E6 and O4 reservists, respectively. The computation for reservists assumes a reservist receives 62 days of basic pay, equal to 48 days of drills plus

Table 2.3
TSP Individual and Agency Automatic and Matching Contribution Rates

Individual Contribution (%)	Agency Automatic Contribution (%)	Agency Matching Contribution (%)	Total TSP Contribution (%)
0	1	0	1
1	1	1	3
2	1	2	5
3	1	3	7
4	1	3.5	8.5
5	1	4	10

SOURCE: DoD, 2017.

14 active duty training days. Consequently, the annual reserve contributions are prorated by 62 ÷ 365.

Continuation Pay

Continuation pay is a new element of compensation under the BRS: a one-time payment that increases current compensation. The purpose of CP is to sustain the size and experience mix of the force by providing a retention incentive to those in their midcareer to offset the reduction in retention incentives caused by the reduced DB multiplier. The TSP also offsets the reduced DB multiplier, but its effect on midcareer retention is muted, as it is not payable until age 59 and a half, while CP is an increase in compensation in the midcareer.

The services have the option of offering CP between YOS 8 and 12. For RC members, YOS is based on number of "good" years, where a good year is completed when a reservist has earned at least 50 points per year so that the year counts toward retirement eligibility. CP would be in addition to any special and incentive pay currently offered to service members. Insofar as members are forward-looking, CP provides an inducement for those with fewer YOS to stay until they can receive CP. Once they reach the YOS when they receive CP, they can receive the CP provided they make a three-year service obligation. In the case of CP offered at YOS 12, by YOS 15, the incentive to stay provided by the availability of the DB annuity at YOS 20 has become relatively strong, and few leave after YOS 15.

Under NDAA 2016, CP was mandated to be paid at YOS 12 and required a four-year service obligation. NDAA 2017 changed CP to allow the services to pay it at any YOS between eight and 12 and reduced the obligation to three years. Our 2017 study of the BRS for the Army and the other services assessed the BRS under NDAA 2016 and did not consider the NDAA 2017 changes. To be consistent with our 2017 analysis of the BRS for the Army, our current analysis for the USAR assumes that the CP multiplier is paid at YOS 12 and requires a four-year obligation, though the model could be modified to allow shorter obligations or different eligibility YOS.

CP is a multiple of monthly basic pay. AC members are guaranteed a minimum of 2.5 months of basic pay (i.e., a 2.5 multiplier), and RC members are guaranteed a minimum of 0.5 months of basic pay (a 0.5 multiplier). A service may pay CP above these minimums, and the service would have to request funds to cover the cost of doing so. The CP multiplier may vary across members and could vary for officer and enlisted personnel and for the active and reserve components. CP entails a three-year service obligation. Members who leave the force before completing their three-year obligation are required to repay CP on a prorated basis. For example, a member who served only one year out of the three would be required to repay two-thirds of CP received at YOS 12.

Table 2.2 provides examples of the computation of CP for an E6 and an O4, assuming the multipliers are set at the floors of 2.5 and 0.5 for the AC and RC, respectively. For the enlisted reservist, CP would be $1,782 and would be $3,527 for the officer. For AC personnel, CP would be $8,910 and $17,633 for the E6 and O4, respectively.

Continuation pay is not intended to replace existing special and incentive pay that target compensation to service members in recognition of differences in working conditions, risk of danger, nature of work, specialized skills, and unusual external civilian opportunities. These kinds of pay are expected to continue as they have in the past. A number of these kinds of pay are intended to help sustain retention in specific occupations, such as medical-related specialties. As the legacy retirement system and basic pay table do not vary with occupation, it is not necessarily the case that CP would vary by occupation, and we do not model CP as varying across personnel within the enlisted or officer force for any of the Army components. That said, the Army does have the discretion to allow CP multipliers that are above the minimum to vary across occupational areas.

As we will discuss in Chapter Four, we consider four CP courses of action. The first COA sets the CP multipliers to their minimum levels mandated by Congress, 2.5 for the RA and 0.5 for the USAR and the ARNG. The second determines the CP multipliers for the RA, USAR, and ARNG at the values producing the best fit to the baseline RA, USAR, and ARNG force sizes and retention profiles (cumulative retention probability by years of service), given the other elements of the reform. This second COA assumes that the USAR would work jointly with the RA and ARNG to set the optimal multiplier to achieve the baseline force sizes and retention profiles. The third COA assumes that the ARNG does not coordinate with the USAR and RA; the CP multipliers for the RA and USAR are the values producing the best fit to the baseline RA and USAR force sizes and retention profiles, ignoring how ARNG sets its CP multiplier. For the purpose of our analysis, we assume the ARNG selects the CP floor in the COA. The fourth COA assumes that the USAR acts independently of the RA and ARNG in setting its CP multiplier, so the CP multiplier is optimized so that it achieves the baseline USAR force size and retention profile. In our analysis of this COA, we assume the RA and ARNG select the mandated congressional floors for CP.

Opt-In

A final feature of the BRS concerns the transition to the new plan. All members serving as of December 31, 2017, are grandfathered under the legacy system, but those with 12 or fewer YOS have the choice to opt in to the new system between January 1, 2018, and December 31, 2018. All new members who enter after January 1, 2018, are automatically enrolled in the new system. In the case of the RC, members must be performing RC service on December 31, 2017, meaning that they must be receiving pay to be eligible to enroll in the BRS. Performing reserve service means the individual is a member of the active ARNG; a member of the active USAR; one of a full-time support personnel; or a member of the selected reserve, individual ready reserve, or active standby reserve. In the case of members of the latter two groups, individuals are eligible for the BRS, but to enroll they must also be receiving pay. If they are not in a paid status during 2018, they have a 30-day window to enroll the first time they return to paid status after 2018. Thus, an eligible member of the individual ready reserve, for example, could potentially enroll in 2019 or some date thereafter.

For RC members, YOS is calculated by total retirement points divided by 360. Consequently, a member with fewer than 12 × 360 = 4,320 points as of December 31, 2017, is eligible to elect the BRS. Because many RC members do not accumulate 360 points in a given year, and indeed only require 50 points per year for that year to count toward retirement eligibility (a "good" year), RC members could be eligible to opt in to the BRS because they have fewer than 4,320 points, but have more "good" years of service in terms of seniority. That is, an RC member could have more than 12 "good" years of service countable toward retirement vesting, but have fewer than 4,320 points.

CHAPTER THREE

Overview of the Dynamic Retention Model, Simulation, and Application to the Blended Retirement System

Dynamic Retention Model Overview

Decisionmakers concerned with force management need answers to questions about how changes to the level and structure of military compensation—such as the BRS—affect retention over a military career. This requires a capability that provides quantitative estimates of how AC retention, RC participation, and costs are affected by changes to present and future compensation. The capability needs to be based on a solid theory of retention decisionmaking over a service member's career, it needs to be empirically grounded in data on actual retention behavior of service members over a long period of time, and it needs a simulation capability that allows us to assess major compensation reforms without relying on the existence of prior variations in such reforms. That is, it needs to be a capability that allows "what if" analyses. The DRM provides these capabilities.

The DRM is a stochastic dynamic programming model that describes an individual's retention decisions over the individual's career. It models service-enlisted personnel and officers as being rational and forward-looking each time they make the retention decision, taking into account both their own preference for military service and uncertainty about future events that may cause them to value military service more or less, relative to civilian life. At each decision point of the military career, the individual compares the value of leaving the military with the value of staying, taking into account that the decision to stay can be revisited at a later time. For example, the value of leaving the AC includes the discounted present value of the stream of income from a civilian career, plus any AC military retirement benefit the individual may have vested in, plus the option value associated with any possible RC career in the USAR or ARNG (RC compensation, intrinsic satisfaction from service, and possibly a retirement benefit). The value of staying in the AC includes any intrinsic benefit individuals receive from military service: their monetary compensation; one more year toward vesting in the AC defined benefit (if they have not already vested) plus the option value of being able to decide whether to stay or leave in the next period, defined as a year in our model. The value of staying thus implicitly includes the discounted present value of any military retirement benefit that may accrue to the individual, weighted by the probability that the individual will qualify for that benefit.

In this project, we extended the DRM that we developed in past studies, mentioned in Chapter One, to distinguish between USAR and ARNG service. Our earlier work only distinguished between AC and RC service, lumping the USAR and ARNG together. In this chapter, we describe the extended version of the DRM. A more formal mathematical rendition of the model is presented in the appendix.

We have two versions of the extended DRM. The first focuses on individuals who begin their military service in the AC.[1] During each period of active service, the individual compares the value of staying in the AC with leaving to become a civilian and possibly participating in the RC through the USAR or ARNG, and bases his or her decision on which alternative has the maximum value. Every year, after leaving active service, the individual compares the value of leading a purely civilian life with the value of participating (or continuing to participate) in either the USAR or ARNG, and chooses among the three

[1] In this report when we refer to AC we mean members of the AC who are also RA; that is, AC should be taken to exclude full-time reservists and any other individuals whose status does not make them part of the RA.

alternatives (civilian, USAR, or ARNG) the one that yields the maximum value for that year. Importantly, members can switch between serving in the USAR and the ARNG.

The second version is a non-prior AC (RA) service RC model that focuses on individuals without prior active service, who begin their military service in the USAR (or, alternatively, in the ARNG). During each period, a member can compare the value of participating in the USAR (or ARNG) with the value of leading a purely civilian life and choose the alternative that yields the maximum value. The structure of the two versions is similar, so we focus on the first version in our description and discuss how the second version differs later in the chapter.

A key feature of the model is that an individual can choose to revisit the decision to stay in the AC or participate in the USAR or ARNG at a later date, and that decision will depend on the individual's unique circumstances at a given point in time. Those circumstances include preferences for RA, USAR, or ARNG service relative to a purely civilian life and random events that may affect relative preferences over AC, civilian, and USAR and ARNG alternatives.

In the model, the value of staying in the AC depends on the individual's preference (or "taste") for active military service unrelated to compensation, which is assumed to be constant over time for an individual; the compensation received for active service; the expected maximum of the value of staying and leaving in the next period; and a period- and individual-specific environmental disturbance term (or "shock") that can either positively or negatively affect the value placed on active service in that period. For example, an unusually good assignment would increase one's relative preference for active service, while having an ailing family member who requires assistance with home care may decrease the value placed on active service. These shocks are unobserved in the data and only reveal themselves via deviations from the expected decisions given observable characteristics. The value of staying also includes the value of the option to leave at a later date; that is, the individual knows that he or she can revisit the decision to stay the next time it is possible to make a retention decision.

We make the simplifying assumption that once individuals have left active service, they do not reenter. While there are instances where people do reenter the AC, well over 90 percent of those who leave do not reenter. This assumption substantially reduces the number of possible career paths that need to be evaluated and makes the model more tractable.

An individual who leaves the AC can choose to either be a civilian or combine civilian life with RC service, where RC service involves participation in either the USAR or ARNG. A person can join the USAR or ARNG immediately after leaving active service, or can choose to join at a later date. Once a person enters the USAR or ARNG, the individual is free to choose to stay or leave, with the option of reentering at a later date or with the option of switching components, service regulations permitting. For example, a USAR member may opt to leave and be a pure civilian for two years and then return to the Army by serving in the ARNG. Or alternatively, a member may switch directly from the USAR to the ARNG with no gap. Figure 3.1 shows how this process works over three periods, and the different decisions facing an individual at each point in time depending on whether the individual is in the AC, USAR, ARNG, or is a civilian not currently participating in the RC.

At the beginning of each year, RC members compare the value of the civilian alternative—that is, leading a purely civilian life for that year—with the maximum of the value of the USAR versus the value

Figure 3.1
Decision Tree for an Army Member over Three Periods

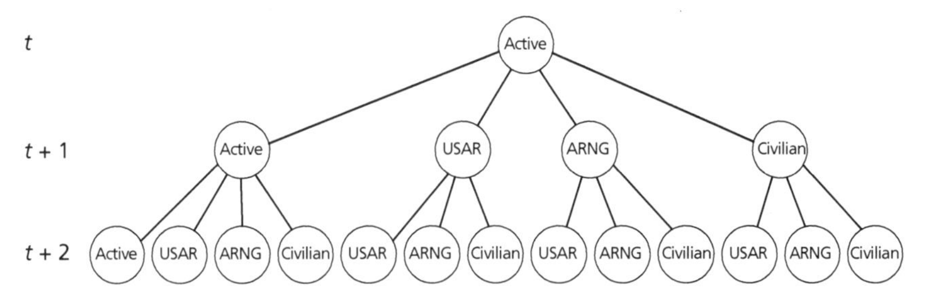

NOTE: t = time.

of ARNG service. That is, the member is assumed to choose to participate in the component that provides the maximum value (USAR versus ARNG), and then given that maximum value, to compare it to the value of the civilian alternative. A member chooses to participate in a first or additional year of RC service by taking the alternative that yields the maximum value.

The value of the civilian alternative includes the civilian wage, the AC or RC military retirement benefit the individual is entitled to receive (if any), an individual- and period-specific shock term that can either positively or negatively affect preference for the civilian alternative, and the future option to enter (or reenter) the USAR or ARNG—service regulations permitting.

The value of RC service includes the civilian wage; the RC compensation to which the individual is entitled, given his or her cumulative AC and RC service; an individual- and period-specific shock term that can either positively or negatively affect the preference for the USAR or ARNG alternative; and the future option to either continue in the RC or return to a purely civilian life. A positive reserve shock or negative civilian shock makes entering the reserves more attractive.[2] Because the model assumes that the reservist is employed in a civilian job, another year in the reserves implies another year of civilian experience (and assumes the reservist is not deployed), so the civilian wage increases too.[3] If the reservist leaves the reserves, military experience does not increase but civilian experience does—and so does the civilian wage.

---

[2] "Shock" refers to a random draw from a distribution representing events that can affect the individual's perceived value of a choice, such as the choice to continue in the active component or leave to become a civilian or reservist. These types of events could include an unusually good or bad assignment, a spouse requiring home caregiving, and so on. This approach is used because events that affect the perceived value of a choice should be recognized in the decision, yet most data sets, including ours, do not contain variables for all the various types of events that could affect these values. Although in reality there may be many such events in any period, we use a single shock (random draw) to represent their net effect. Thus, the random draw can be positive or negative, big or small.

[3] Staying in the reserves might alter civilian options upon leaving the reserves; for example, the reservist might acquire skills that are valuable for civilian jobs. We do not model this possibility.

As mentioned, the model assumes individuals differ in their tastes for AC, USAR, and ARNG service, and individuals with different tastes may respond differently to the same policy. Parameters estimated from empirical data include those for the multivariate distribution of tastes over members at entry (i.e., means, standard deviations, and correlations) and the environmental shock distributions (i.e., location and scale terms). The model also embeds details of the compensation and retirement systems. When the model parameters have been estimated, the model can be used to simulate alternative military compensation and retirement policies. The first challenge that arises in estimating the model is that, while we allow tastes to differ, tastes are not directly observed. Each individual knows his or her own taste and makes decisions based on that taste, but that individual's specific taste is unknown to us. The second challenge arises because we do not observe the random shocks facing members each period (such as an ailing family member). Each individual knows the shock that he or she is currently subject to and makes a decision depending on the nature of this shock, but the individual's current shock is unknown to us as analysts. Future shocks are unknown to the individual and the analyst alike. We address the second challenge by assuming the shock terms have an extreme-value type I probability distribution, which allows us to derive a closed-form solution for the expected value of the maximum given an individual's taste for the AC, USAR and ARNG, and thus a closed-form solution for the probability of staying in the AC or leaving and becoming a civilian, or for choosing to participate in the USAR or ARNG when one is a civilian. We address the first challenge by assuming the tastes for the AC, USAR, and ARNG have a joint trivariate normal distribution, and by calculating the expected value of the probability that a member makes a given choice at a point in time by integrating over the taste distribution.[4] As a result, the probability expressions depend only on the parameters of the shock and

[4] Specifically, we numerically integrate out heterogeneity in taste. For trial values of the taste distribution parameters, possible tastes for the individual are drawn from the distribution. For each taste draw, the career probability expression is evaluated and an average of those probabilities is taken, where the weight on a probability depends on the probability of the drawn taste.

taste distributions, and not on the actual taste or shock terms faced by particular individuals at a particular point in time. We can then write an expression for the likelihood of a given career in terms of the parameters of the normal distribution and the extreme-value distribution. Then, given our data, we use maximum likelihood methods to find the value of these parameters together with the personal discount factor that maximizes the likelihood of observing the observed data.

In summary, the model portrays AC, USAR, and ARNG retention choices as the result of individual members solving a stochastic dynamic program that embeds uncertainty about future conditions (shocks), and permits individual members' tastes for active, USAR, and ARNG service to be unique. We as analysts do not observe these tastes, but we assume they follow a certain joint probability distribution. The model builds in uncertainty in the form of random shocks, and we assume the shocks follow certain probability distributions. We use the model's structure, the distributional assumptions about tastes and shocks, and information about military pay, military retirement benefits, and civilian pay to derive expressions for the probability of a military career represented by AC retention and possible USAR and ARNG participation by period. For instance, a person might serve five years in the AC, then three years in the USAR, then work as a civilian for two years followed by a year in the ARNG, finally settling as a civilian in all remaining periods. Such a career would look like this: {A,A,A,A,A,U,U,U,C,C,N,C,C,...,C} where A represents active service, U represents USAR service, N represents ARNG service, and C represents pure civilian work. We have longitudinal data and create probability expressions for each person's career. These career probabilities are multiplied together to obtain a maximum likelihood expression for the entire sample, and this expression is maximized with respect to the model parameters to obtain parameter estimates.

Data

Our main data file is the Work Experience file (WEX), a Defense Manpower Data Center (DMDC) file that contains person-specific longitudinal records of active and reserve service.[5] DMDC creates WEX data from the active-duty and reserve master files. DMDC uses these files to build a snapshot of all personnel for each reporting period—that includes demographic and work experience information. To maintain the file, DMDC compares data for the current and previous periods and creates three types of records: a gain record, a loss record, and a change record. A gain record is created when a service member's Social Security number (SSN) does not appear in the previous period, but does appear in the current one. A loss record is created when an SSN appears in the previous period but not in the current one. When a loss occurs, all related work experience records are moved to a loss file; these are retrieved when an individual reenters service (i.e., when a gain occurs). A change record is created when there is a change in any of seven variables: service or component, pay grade, reserve category code, primary service occupation code, secondary service occupation code, duty service occupation code, or unit identification code. The WEX record also includes a member's age and gender.

WEX data begin with service members in the AC or RC on or after September 30, 1990. Our analysis file includes Regular Army AC, USAR, and ARNG non-prior-service entrants in 1990 and 1991 followed through 2015, providing 26 years of data on 1990 entrants and 25 years on 1991 entrants. We drew a sample of 216,870 individuals (4,980 officer and 110,876 enlisted AC accessions, plus 101,014 enlisted RC USAR or ARNG accessions) who entered in 1990 or 1991; constructed each service member's history of AC retention, USAR, or ARNG participation; and used these records in estimating the model. We use WEX variables to identify an individual's component (e.g., AC Army, USAR, ARNG) by year from the date of entry onward. We use pay entry base date and component/branch in counting years of AC service and years of USAR or ARNG participation following AC service. We exclude reservists participating as individual mobilization augmentees (i.e., reserve soldiers trained and preassigned to an active organization to meet personnel requirements in the event of mobiliza-

[5] WEX is used primarily for production of Verification of Military Training and Experience DD Form 2586 documents.

tion; this permits rapid expansion of the force in the event of a military contingency).

We constructed data files for enlisted personnel and officers. In constructing the officer data file, we exclude medical personnel and members of the legal and chaplain corps because their career patterns differ markedly from those of the rest of the officer corps. Analysis of retention for these personnel needs to be conducted separately. Furthermore, because there are so few non–prior-service officers participating in the USAR, we did not estimate a non–prior-service model for USAR officers but estimate a non–prior-service model for USAR enlisted personnel only.

The other key data pertain to military and civilian pay. AC pay, RC pay, and civilian pay are averages based on years of AC, RC, and total experience, respectively. AC and RC pay are also related to military retirement benefits. We use data from 2007 for these pays and benefits and then put them in 2017 dollars using the Consumer Price Index for Urban Consumers. Specifically, annual military pay for AC members is represented by regular military compensation (RMC) for FY 2007, which is equal to the sum of basic pay, basic allowance for subsistence (BAS), basic allowance for housing (BAH), and the federal income tax advantage accruing to members because allowances are not taxed. (The tax advantage is computed by finding the additional amount in taxable cash income members would have to receive to end up with the same after-tax income if allowances were taxable; this amounts to 6 percent of pay on average.) We compute RMC by year of service for enlisted personnel and officers using the tables provided by the Office of the Secretary of Defense, Personnel and Readiness, Directorate of Compensation in *Selected Military Compensation Tables* and weighting them with the 2007 grade-by-YOS inventory of enlisted and officer personnel in each service.[6]

RC members are paid differently than AC members even though the same pay tables are used for both AC and RC. Reservists who are

[6] Office of the Under Secretary of Defense, Personnel and Readiness, Directorate of Compensation, *Selected Military Compensation Tables*, Washington, D.C.: U.S. Department of Defense, 2007.

drilling, but not on active duty, receive a subsistence allowance for their two drilling days per month and do not receive a housing allowance. Reservists on active-duty training typically receive rations and housing in kind during the two weeks of training and receive either a partial housing allowance or a rate applied for married members, unless they are housed in contract housing off-base.

RC pay is based on years of AC service and years of RC participation in either the USAR or ARNG. We average RC pay over pay grade and dependency status using RC strength information from the 2007 *Official Guard and Reserve Manpower Strengths and Statistics Report* (Office of the Assistant Secretary of Defense, Reserve Affairs, 2007). Reserve pay in a year is calculated as the sum of drill pay for four drills per month, 12 times a year, plus pay for 14 days of active-duty training. Drill pay is 1/30th of monthly basic pay for each drill period, or 4/30th per weekend. During each day of active-duty training, the reservist receives basic pay plus BAS. Single members receive BAH for a service member without dependents, while married members receive BAH for a service member with dependents. In our calculation, RC members receive BAH RC/T (reserve component/transit), a housing allowance for certain circumstances, including being on active duty less than 30 days. Given years of service and grade, we compute a reservist's annual pay as

12 × weekend drill pay
+ 14 × (BAS + daily basic pay

+ % married × % on base × BAH RC/T for those with dependents
+ % single × % on base × BAH RC/T for those without dependents)

+ tax advantage.

To incorporate the tax advantage, we use the same adjustment as for AC annual pay, 6 percent. Some reservists receive special pay and incentive pay such as bonuses, but these are not included explicitly. Instead, their role is generally picked up in estimated mean of the reserve taste distribution. Also, the model does not address the activation and deployment of reservists.

The reserve retirement benefit formula and the high-three active duty retirement formula are programmed into our model. For the computation of reserve retirement benefits, we assume that an RC participant accumulates 75 points per year. Unlike AC retirement benefits, which start as soon as the AC member retires from service, RC retirement benefits begin at age 60.[7] The formula for RC retirement benefits under the legacy system is the same as that for AC retirement benefits, with the proviso that RC retirement points are converted into years of service (for the purpose of retirement) by dividing total points by 360. A year of AC service counts as a full year. Reservists who qualify for reserve retirement benefits can transfer to the "retired reserve," which means that their high-three pay is based on the basic pay table in place on their 60th birthday, and their basic pay is based on their pay grade and years in grade, where the latter include years in the retired reserve.[8]

For enlisted personnel, civilian earnings are the 2007 median wage by experience for full-time male workers with an associate's degree. For officers, we use the 2007 80th percentile of earnings for full-time male workers with a master's degree in management occupations.[9] The data are from the Census Bureau. Civilian work experience is defined as the sum of active years, reserve years, and civilian years since age 20, but in this case does not vary by other factors such as years since leaving

---

[7] If the RC member has been deployed in the period beginning on January 28, 2008, retirement age is decreased by three months for every 90 consecutive days of deployment. This change is not included in our model because the model does not include deployment.

[8] In addition, military retirees (including reserve retirees receiving retired pay) are eligible to receive health care through TRICARE for the remainder of their lives (as are their spouses) and coverage continues for the spouse if the retiree dies and she or he does not remarry. "Gray area" retirees (i.e., members of the retired reserve who are not drawing retired pay) may purchase TRICARE coverage under the TRICARE Retired Reserve program until they become eligible for TRICARE. We do not model the health benefit, however.

[9] The choice of the 80th percentile was informed by the finding of the 11th Quadrennial Review of Military Compensation that compensation for officers in 2009 was at the 83rd percentile of civilians with equivalent education and experience.

active duty. As with military pay, civilian pay is converted to 2017 dollars using the Consumer Price Index for urban consumers.

Estimation

The prior AC service model has 17 parameters: nine for the distribution of tastes over members at entry, three for the shock distributions, and five switching costs.

The taste distribution is multivariate normal, with six parameters giving the means and standard deviations of AC, ARNG, and USAR taste, and three correlation parameters: AC and ARNG taste, AC and USAR taste, and ARNG and USAR taste.

The three shock distribution scale parameters can be thought of as corresponding to three levels of choices: whether to continue in the AC or not, whether to lead a purely civilian life or participate in the RC, and whether to participate in the USAR or the ARNG. More formally, we use a nested logit approach to capture these decisions, where the active service member is modeled as comparing active service with a civilian or RC nest and the RC participation decision is modeled as comparing civilian opportunities with a USAR or ARNG nest. We assume the location parameter is zero in all the shock distributions.

In addition, we estimate five parameters for switching costs. These reflect the cost associated with switching from one state to another, e.g., from AC to civilian, or from AC to ARNG.

The non-prior-service USAR model is much simpler with five parameters. The parameters are the mean and standard deviation of the taste for USAR service; the scale parameter for the USAR shock; a switching cost, reflecting the cost associated with switching back into the USAR once a member has left possibly due to lack of billets in a local area; and a switching cost of leaving the USAR before the end of the initial service obligation at YOS 6. The ARNG model of non-prior-service entrants is constructed in a parallel fashion, also with the same five parameters.

The models are estimated using maximum likelihood. Writing down the likelihood function requires us to compute the probability of a given career, where a career consists of a sequence of active, civilian, and reserve (USAR or ARNG) states. This computation, in turn, requires us to compute the probability of choosing each alternative in each time period. Given our assumption of an extreme-value distribution for the shock terms, we can solve the dynamic program given values for active, USAR, and ARNG tastes in the case of our prior-service model. The solution gives closed-form solutions for the probability of choosing each of the two or three alternatives available at any given time, and as mentioned we use these to construct a career probability for each individual. The expression for the career probability implicitly depends on the parameters to be estimated (e.g., mean active taste, mean reserve taste, discount rate, and so forth). Because tastes are not known at the individual level, we numerically integrate out heterogeneity in taste.[10]

Standard errors are computed using numerical differentiation of the likelihood function at the parameter estimates to produce the matrix of second derivatives, the Hessian matrix. The standard errors are computed using the square root of the absolute values of the diagonal of the inverse of the Hessian. We report the parameter estimates and standard errors for officers and enlisted personnel in the appendix.

To judge goodness of fit, we use the parameter estimates to simulate AC, USAR, and ARNG retention patterns by year of service. These simulations are compared to the actual data to assess the extent to which the model predicts actual behavior (i.e., to assess model fit). We also simulate AC, USAR, and ARNG retention under the alternative policies under consideration. Later in this chapter, we describe our simulation approach. Results of simulating the policy proposals are presented in the next chapter.

[10] As explained in an earlier footnote, for trial values of the taste distribution parameters, possible triplets (three-tuples) of tastes for the individual are drawn from the distribution. For each taste draw, the career probability expression is evaluated and a weighted average of those probabilities is taken, where the weight depends on the probability of the drawn taste. Numerical optimization of maximum likelihood is done using a standard Broyden-Fletcher-Goldfarb-Shanno hill-climbing algorithm.

Model Fits

Figure 3.2 shows the fit for the two models of RC participation by individuals with no prior AC service. The left graph shows the fit for the USAR model, while the right graph shows the fit for the ARNG model. The vertical axes show end strength relative to the initial population at the first year of service, and the horizontal axes show cumulative years of RC service. The observed cumulative retention over time is shown via a black Kaplan-Meier survival curve, and the simulated cumulative retention over time is shown via a red survival curve. In both cases, the simulated curve hews closely to the observed curve. Both models slightly underpredict retention in the early years of service (corresponding to the first term), slightly overpredict toward the early midcareer, and follow the observed curve in later years.

Figures 3.3 and 3.4 show the model fits for the enlisted personnel and the officers with prior AC service. Unlike the two independently estimated models of enlisted personnel with no prior AC service shown in Figure 3.2, the prior AC service USAR and ARNG participation curves are the result of a single model where individuals start in the

Figure 3.2
Model Fit for Army USAR and ARNG Enlisted Personnel with No Prior AC Service

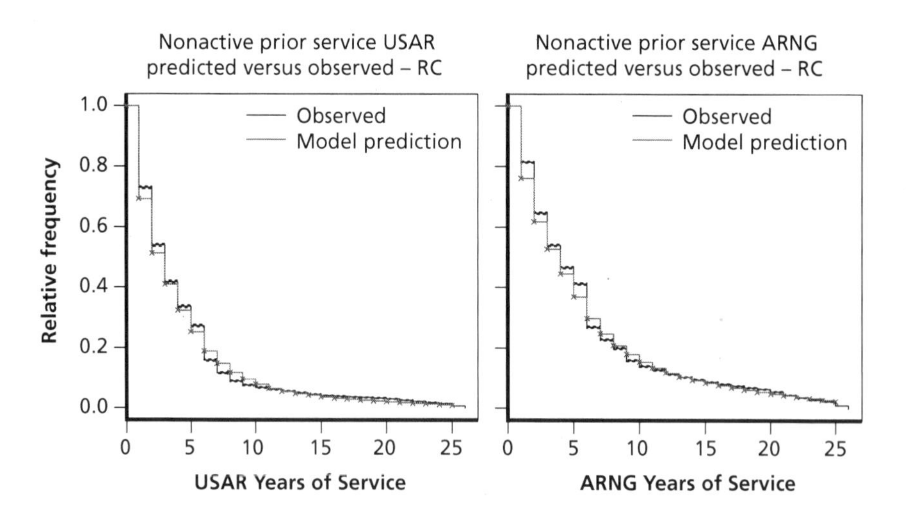

AC, and then can choose later to participate in either the USAR or the ARNG. The curves look very different compared with the ones shown in Figure 3.2, as individuals can enter either the USAR or the ARNG with seniority from prior AC service.

Turning first to the model of enlisted personnel participation in Figure 3.3, we see that the simulated force shape replicates the major features of the observed cohorts, with fit being closer in the USAR graph than in the ARNG graph. In particular, the model underpredicts ARNG participation in AR YOS 4 though YOS 9 where AR refers to active plus reserve years. On the other hand, the USAR-predicted force shape is closer to that observed, less substantially over- and underpredicting retention over the same interval. In both cases, the model does less well than the no prior AC service models do in replicating the behavior of more senior members of the force. That said, by capturing the main features of the profile and providing a good fit at YOS 12 when we assume CP is paid, we expect that the simulation results will not be qualitatively affected by the fit issue.

Figure 3.3
Model Fit for Army USAR and ARNG Enlisted Personnel with Prior AC Service

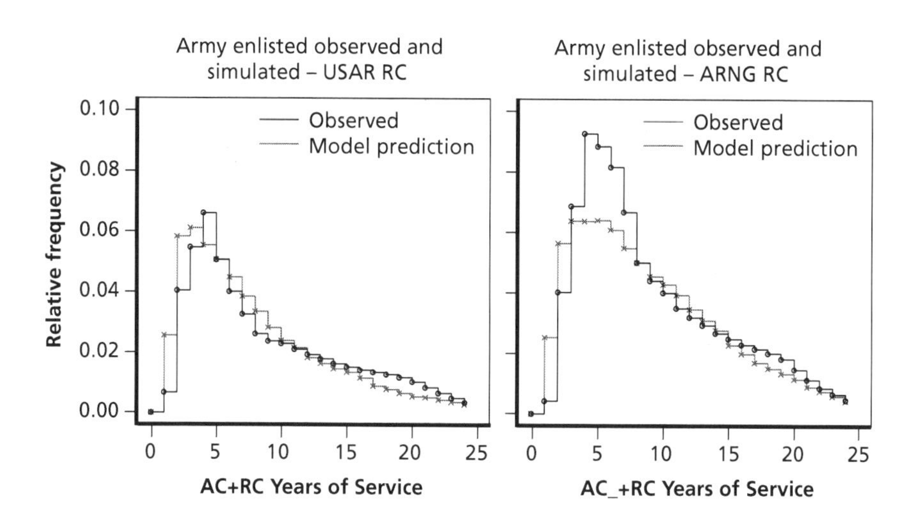

Figure 3.4
Model Fit for Army USAR and ARNG Officers with Prior AC Service

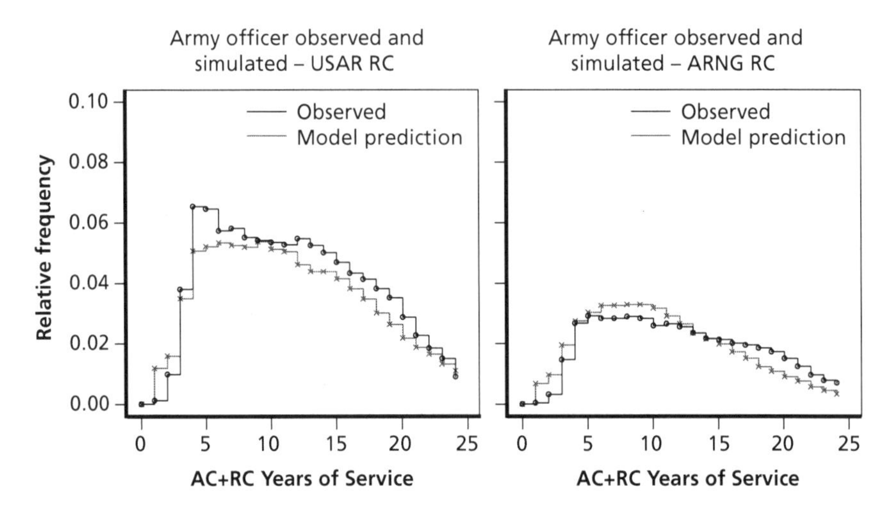

The model of USAR and ARNG officers with prior AC service in Figure 3.4 performs almost equally well between the USAR and ARNG, with perhaps a slight edge going to the ARNG fit. In both cases the major underlying features of the observed data are replicated. The model overpredicts participation in the USAR in the first three years, and underpredicts thereafter, while participation in the ARNG is overpredicted in the early years, underpredicted in the midcareer, and overpredicted in the senior years, so in some sense is more "centered" over the observed data.

Finally, although our focus is on the USAR, we jointly model RA retention and USAR and ARNG participation. Consequently, we also have model fits for RA enlisted personnel and officers. These are shown in Figure 3.5. As in our earlier work, the model fits are very good.

Approach to Simulation and Application to the Blended Retirement System

To simulate retention behavior for the first version of the model that assumes all members start military service entering the AC, we first create a synthetic population of 10,000 individuals entering active duty by randomly drawing tastes from the estimated AC/USAR/ARNG taste distribution. This synthetic population is large enough for the simulations to produce AC retention and USAR participation careers that, when aggregated, provide AC cumulative retention and USAR participation curves representative of the policy being simulated. Each triplet of AC, USAR, and ARNG taste draws represents an individual entering active duty. We also draw shocks for each year for each synthetic individual from the shock distributions. We assume that the synthetic individuals follow the logic of the model, and we specify the

Figure 3.5
Model Fit for RA Officers and Enlisted Personnel

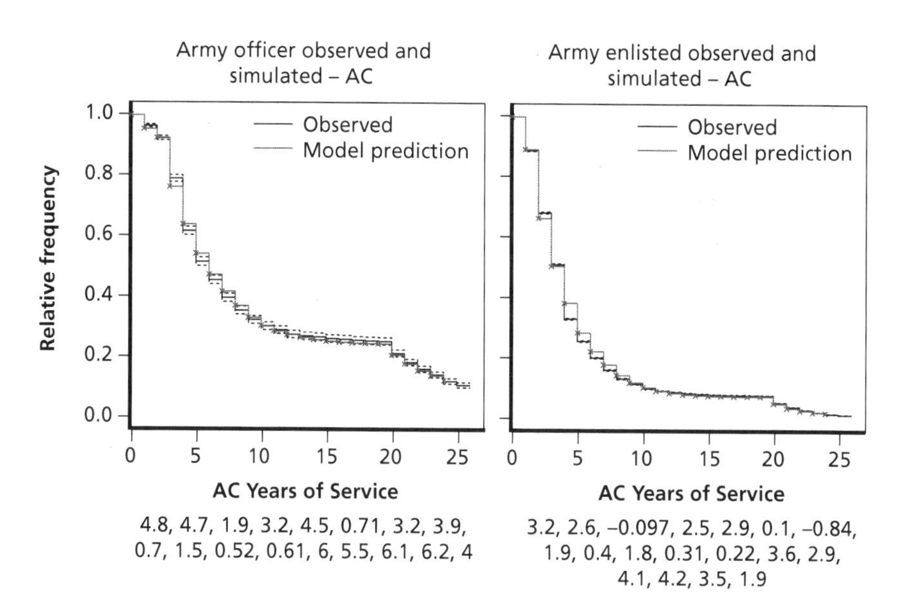

compensation policy for the simulation. We simulate behavior under the legacy retirement system, the baseline, and then simulate it under the BRS. The simulations produce a 30-year record of AC retention and USAR and ARNG participation for each member of the synthetic population under each retirement policy.

Similarly, for the model of non-prior-service entrants to the USAR (and similarly for the ARNG), we create a synthetic population of 10,000 individuals entering USAR by randomly drawing taste from the estimated USAR taste distribution. We also draw shocks for each year for each synthetic individual from the shock distributions. We assume that the synthetic individuals follow the logic of the model for non-prior-service personnel, and we specify the compensation policy for the simulation. We simulate behavior under the legacy retirement system, the baseline, and then simulate it under the BRS. The simulations produce a 30-year record of USAR participation for each member of the synthetic population under each retirement policy.

We use the datasets of simulated behavior to tabulate AC, USAR, and ARNG retention and, along with information on compensation, to compute policy cost and specifically the cost of continuation pay under the BRS. The simulation outputs include graphs of AC retention by year of service and USAR and ARNG participation by year of active-plus-reserve service, as well as tabulations for AC, USAR, and ARNG force sizes and CP costs.

Under the assumption of a steady state, the AC force size of the simulated population is the sum of individuals present in each YOS. This count is scaled up to the 2009 AC force size for Army officers and enlisted personnel, equal to 90,795 and 458,220, respectively.

USAR and ARNG force sizes of members with prior active service are based on the count of simulated individuals participating in the USAR and ARNG at each year of service, given the scaling up of the AC force to its force size. As mentioned, USAR and ARNG YOS are based on the number of active years plus reserve years.[11] For

[11] As an example of this count, consider someone who over the course of 40 years (ages 20 to 60) had 5 years of AC and 5 years of USAR service. This individual would be present in the USAR at six YOS (5+1), 7 (5+2), 8, 9, and 10 (participation in the USAR could have occurred

USAR and ARNG force sizes of members with no prior service, we use 88,000 for the USAR and 223,400 for the ARNG for FY 2009.[12]

Optimization

Continuation pay for the AC and RC equals a continuation pay multiplier times the active-duty monthly basic pay. In our simulations of the BRS, we assume CP is paid at YOS 12, with a pay-back feature for those who separate before completing four additional YOS. In the next chapter, we discuss alternative courses of actions for setting CP. Several of these alternatives involve setting the CP multiplier so that the Army sustains the baseline retention profiles under the legacy system for enlisted personnel and officers.

To achieve this in our model, our simulations compute optimized values of the CP multiplier, given the other features of the BRS. This involves computing the value of the multiplier that minimizes the difference between the baseline retention profile under the legacy high-three retirement system and the profile under the BRS, given the other features of the BRS, including our assumed TSP contribution rate and annuity choice.

A key question is, what is the relevant baseline? Ideally, the baseline retention profile should reflect the service's required experience mix and force. The baseline we use is the simulated retention profile under the current compensation system and high-three retirement system. While there is no presumption that future requirements will call for the same size and mix as the baseline force, we assessed the retention effects of the BRS in terms of how well it could achieve the baseline.

---

in nonconsecutive calendar years). In each of these years, the individual would be counted in the steady-state USAR force. Because everyone begins in the AC, the smallest USAR YOS entry is 2 (1+1).

[12] These figures are estimates based on the number of non-prior-service enlisted personnel gains and total enlisted personnel for the USAR and ARNG for FY 2009, respectively, from the 2009 Population Representation of the Armed Services (Tables C-1 and C-11) together with DRM retention estimates for the prior-service populations. See Office of the Under Secretary of Defense, Personnel and Readiness, 2009.

We show the optimized multipliers in the next chapter, where we discuss steady-state results.

Simulation of Continuation Pay Cost

Continuation pay costs for the AC equal CP costs at YOS 12 multiplied by the number of AC personnel at YOS 12 and the number of USAR personnel at YOS 12. CP costs are scaled up to the 2009 force size and expressed in 2016 dollars.

Simulation of Opt-In Behavior

We use the mathematical structure of the DRM to simulate the choice between the BRS and the legacy system for those who would be eligible. In the DRM structure, we compute the expected value of staying in the RA, USAR, or ARNG at each point in a member's career. We use this computation and assume members would elect to opt in if the value of staying in the component (RA, USAR, or ARNG) at the time of the choice was higher under the new system than under the legacy system. This allowed us to compute the percentage of members in a component who would opt in by years of service at the time of the opt-in decision. This approach takes advantage of the facts that the opt-in choice is offered to incumbent members of the RA, USAR, and ARNG, and, because a member can remain under the legacy system, the choice to opt in cannot make a member worse off but only offers the possibility of being better off. These facts mean that the opt-in choice (a) is not relevant to individuals who are not already a member of the RA, USAR, or ARNG, (b) has no chance of inducing an incumbent member to leave military service, and (c) might make the member better off financially.

Limitations, Advantages, and Assumptions

The DRM has several limitations. The model assumes that real military pay, promotion policy, and real civilian pay do not vary over time, and it excludes demographic factors such as gender, marital status, and spousal employment. It excludes health status and health care benefits, and we do not model deployment or deployment-related pay. The model also assumes that the Army manages selective reenlistment bonuses and other special pay and incentive pay, increasing and decreasing as retention changes, so that retention flows are relatively stable over time, even when end strength is changing. That said, the estimated models fit the observed data reasonably well for Army officers and enlisted personnel in the RA, USAR, and ARNG.

It is also important to recognize some limitations of our modeling that are specific to simulating the BRS reform. The DRM does not model members' choices regarding an annuity or lump sum for AC or RC members. The DRM also does not model members' savings decisions, and therefore their decisions regarding whether and how much to contribute to the TSP. Due to this, we are not able to simulate what percentage of members will choose a full annuity versus a partial lump sum or a full lump sum, nor are we able to simulate the distribution of contribution rates among service members to the TSP and, therefore, the average TSP match rate.[13]

We manage these limitations by assigning all members the same assumed choice in the simulation. In the case of the TSP contribution rate, we assume members contribute 5 percent of their basic pay, thereby receiving the full 4-percent DoD match rate, on top of the 1-percent automatic contribution.[14] We used the same assumption in our earlier analysis of the BRS for the other services. We also used the same assumption in our earlier analysis with respect to the lump-sum choice. Here, as there, we assume all members choose the annuity and do not choose a lump sum.

---

[13] The DRM could be extended to include these decisions. However, data are currently not available to allow an empirical implementation of this extension.

[14] Prior analysis showed that retention effects were similar under the assumption that all members contributed a lower percentage, e.g., 3 percent instead of 5 percent, and with optimized CP multipliers. With a 5-percent contribution, the member realizes the greatest gain from the TSP, and the incentive to do this is strong because of dollar-for-dollar matching up to 3 percent and half dollar-for-dollar matching at 4 and 5 percent. With higher TSP contributions from the service, the optimal CP multiplier is slightly lower (Asch, Mattock, and Hosek, 2015).

Having a choice of a lump sum or annuity is a valuable feature of the reform package. Similarly, the availability of a DoD matching contribution is a valuable feature. This additional value improves the value of staying in the military and therefore improves retention. However, data are not yet available for us to extend the DRM to include these choices and estimate parameters pertaining to them. As a result, we cannot quantify the added value of having the choices or incorporate the added value in our simulations. Given that the value of these choices is omitted, we understate the retention effect of the reform package by an unknown amount.[15]

---

[15] Further, were the value of these choices included, the optimized CP multiplier probably would be somewhat smaller.

CHAPTER FOUR

Courses of Action and Steady-State Results

The OCAR requested that we consider alternative COAs for setting CP under the BRS for the USAR. We begin the chapter with a discussion of the COAs that we considered. We then show the effects of the BRS on steady-state RA retention and USAR and ARNG participation under the alternative COAs. By steady state, we mean when all members have spent an entire career under the BRS. We conclude with estimates of the steady-state CP costs under the alternatives.

Courses of Action for Continuation Pay

The first course of action we consider is the case when the CP multipliers are set at the floors stipulated by Congress. As discussed in Chapter Two, the floor for the AC is 2.5 and is 0.5 for the RC. Table 4.1 shows the COAs we consider and the optimized CP multipliers under each COA. It also shows that the first COA is setting CP at the floor for officers and enlisted personnel.

The other courses of action are guided by our earlier analysis of the BRS for all five armed services (Army, Navy, Marine Corps, Air Force, and Coast Guard) where we optimized the CP multipliers for enlisted personnel and for officers in each service in order to sustain retention under the BRS relative to the baseline, defined as retention under the legacy retirement system (Asch, Mattock, and Hosek, 2017). In the previous analysis of the BRS, we found that optimized multipliers for enlisted personnel in each service were about 2.5 for AC personnel and about 0.5 for RC personnel. For officers, the optimized mul-

Table 4.1
Continuation Pay Courses of Action and Optimized CP Multipliers

COA	RA	USAR	ARNG
Enlisted personnel[a]			
1. RA, USAR, ARNG set to floor	2.50	0.50	0.50
Officer			
1. RA, USAR, ARNG set to floor	2.50	0.50	0.50
2. Optimize RA, USAR, ARNG	8.28	2.25	1.08
3. Optimize RA, USAR; ARNG set to floor	7.59	1.78	0.50
4. Optimize USAR; RA and ARNG set to floor	2.50	0.19	0.50

NOTE: Multipliers represent months of basic pay at YOS 12.
[a] COAs 2–3 are not relevant for enlisted personnel, because retention is sustained under COA 1.

tipliers were about 12 for AC personnel, or about a year of basic pay, and about six months or half a year of basic pay for RC personnel. The optimized multipliers were quite similar across the services. In the case of the Army, we found that the optimized CP multipliers for AC and RC enlisted personnel were 2.39 and 0.45, respectively, and for officers were 10.85 and 5.62, respectively. That is, the optimized multipliers for Army enlisted personnel are close to the congressional floor, but for officers, the multipliers that sustain officer retention are substantially higher.

In our current analysis, we considered three additional alternative courses of action that involved optimizing CP multipliers, COAs 2–4, as shown in Table 4.1. Note that we only consider these additional COAs for officers; the CP multiplier floors sustain RA, USAR, and ARNG enlisted retention in the steady state, as we show in the next subsection, so the optimized CP multipliers for enlisted personnel will equal the floors. As a result, there is no need to consider additional COAs for them. But, this is not the case for officers. As we show in the next subsection, and as we found in our earlier analysis, since the floors do not sustain retention for officers, we consider COAs that would sustain officer retention under the BRS relative to the legacy system.

The COAs reflect different possible scenarios regarding the level of cooperation between the Army components in setting the CP multipliers. With full cooperation between the components in the setting of CP, it is assumed that the multiplier for the USAR will be jointly optimized to also sustain RA retention and ARNG participation. This represents COA 2. COA 3 captures the scenario where the USAR and RA work together to jointly set the CP multipliers in those components to sustain retention, but the ARNG operates independently, perhaps in coordination with the states or with the Air National Guard. Finally, COA 4 represents the scenario where the USAR operates independently of both the RA and the ARNG. Analysis of these scenarios requires an assumption about how components behave in setting CP if they are not working with the other components. For simplicity, we assume that they set CP at the congressional floor.[1]

More specifically, for COA 2, we jointly optimize the RA, USAR, and ARNG CP multipliers to sustain RA retention and USAR and ARNG participation under the BRS, given the other features of the BRS, relative to the baseline which is assumed to be the legacy high-three retirement system. Jointly optimizing the multipliers explicitly recognizes that the CP multiplier in the AC will affect not only retention in the AC but the flow of personnel with AC service to the USAR and ARNG as well. Similarly, it recognizes that the setting of the CP multiplier in the USAR will affect RA retention and participation in the ARNG. By explicitly recognizing the intercomponent flows and choosing CP multipliers that sustain retention in all three Army components, COA 2 takes the perspective of overall Army leadership, rather than leadership of the RA, USAR, or ARNG separately.

As shown in Table 4.1, the jointly optimized CP multipliers for RA, USAR, and ARNG officers under COA 2 are 8.28, 2.25, and

---

1 Other possibilities could be considered. For example, a game theoretic approach could be taken where each component operates strategically and sets CP in anticipation of what the other components might do.

1.08, respectively. Interestingly, the optimized CP multiplier for the ARNG is lower than the one for the USAR.

COA 3 assumes that the Army jointly optimizes the CP multipliers for the RA and USAR, but that the ARNG acts independently. For the purpose of our analysis, we assume the ARNG sets the officer CP multiplier to the floor of 0.5. In this case, we find that optimized multipliers for RA and USAR officers are 7.59 and 1.78, respectively, a bit lower than the optimized multipliers in COA 2. This suggests that ignoring whether their CP multiplier policy affects ARNG participation has some effect on the best multipliers for the USAR and RA in terms of sustaining retention.

For COA 4, we assume that the USAR chooses the CP multiplier that sustains USAR participation, independent from the CP multiplier choice taken by the AC or the ARNG. For the purpose of our analysis, we assume the RA and ARNG choose the mandated floors of 2.5 and 0.5, respectively. The optimization methodology is described in the previous chapter, and we find that under COA 2, the optimized CP multiplier for USAR officers is 0.19, below the 0.5 mandated floor. Because the optimized CP multiplier is below the floor, this COA is not feasible and we do not show results for this COA.[2]

Simulated Steady-State Effects of the Blended Retirement System on Active Component Retention and Reserve Component Participation

Figures 4.1–4.4 show the results for enlisted personnel and officers for COA 1 where the CP multiplier is set to the congressionally mandated

---

[2] Three additional COAs that optimize CP multipliers could be considered:

- optimizing the RA multiplier alone, ignoring the USAR and ARNG (e.g., assuming they act independently)
- optimizing the RA and ARNG multipliers jointly, ignoring the USAR
- optimizing the ARNG multiplier alone, ignoring the RA and USAR.

We did not consider these other COAs because the focus of this report's sponsor is on the USAR specifically, so the COAs with optimized CP multipliers that we considered all involved sustaining USAR participation.

floor of 2.5 months and 0.5 months for AC and RC personnel, respectively. Figure 4.1 shows results for the USAR and ARNG for enlisted personnel with no prior service, while Figures 4.2 and 4.3 show results for enlisted personnel and officers with prior AC service, respectively. Figure 4.4 shows results for enlisted and officer RA personnel. Each figure shows the simulated retention profile by YOS under the legacy baseline retirement system (black line) versus the BRS (red line).

Consistent with our earlier analysis of the BRS (Asch, Mattock, and Hosek, 2017), our simulations show that the 2.5 and 0.5 multiplier floors, together with the other elements of the BRS—a lower DB multiplier of 2.0 percent and the addition of the TSP—would sustain enlisted RA retention and USAR and ARNG participation in the steady state relative to the baseline for both prior- and no-prior-service personnel. The red and black lines are virtually indistinguishable for Army enlisted personnel.

Table 4.2 shows the percentage change in RA, USAR, and ARNG force size relative to the baseline under each COA. Under COA 1 for

Figure 4.1
COA 1: Participation of USAR and ARNG Enlisted Personnel with No Prior Service Under the BRS Versus Baseline

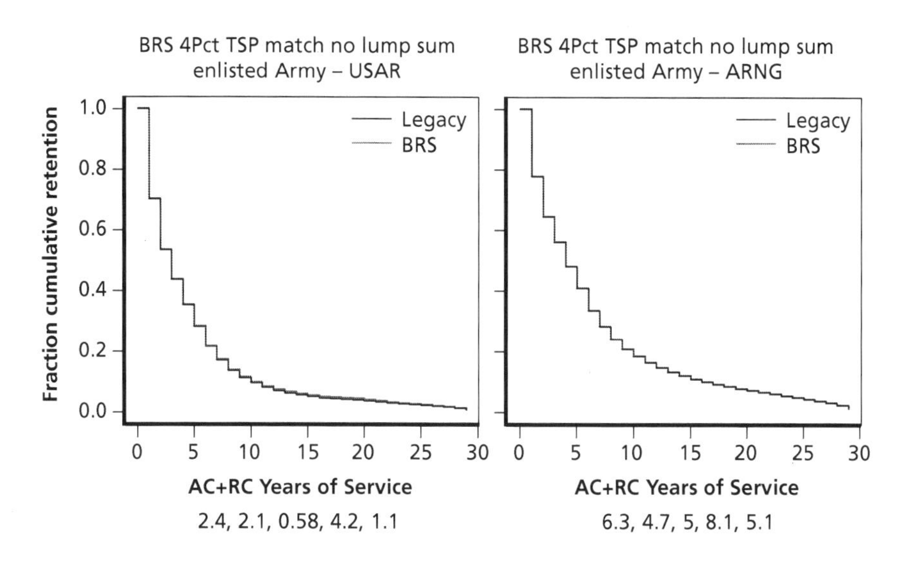

Figure 4.2
COA 1: Participation of USAR and ARNG Enlisted Personnel with Prior RA Service Under the BRS Versus Baseline

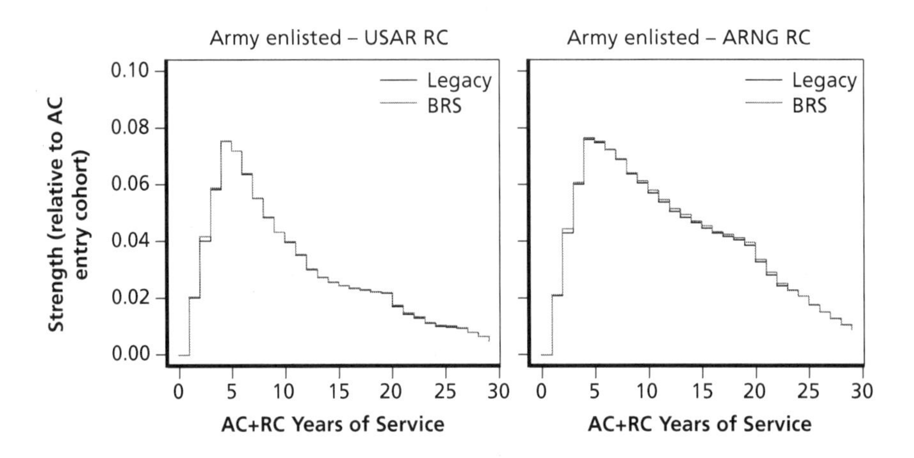

Figure 4.3
COA 1: Participation of USAR and ARNG Officers with Prior RA Service Under the BRS Versus Baseline

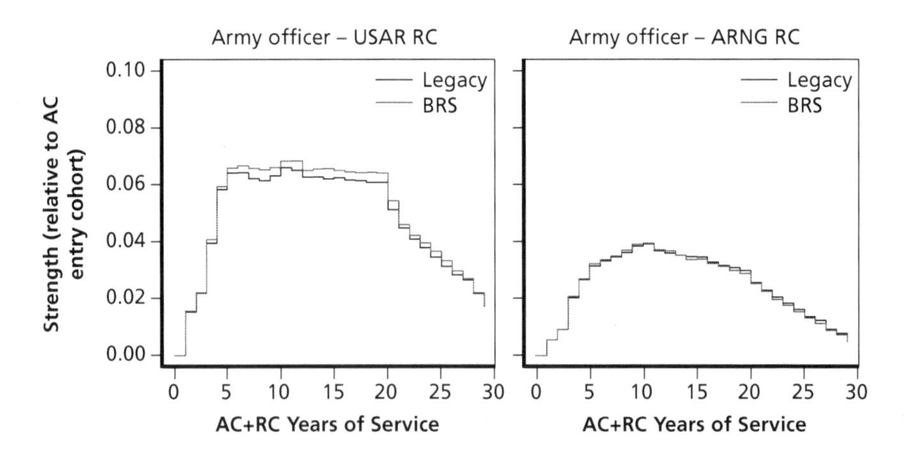

enlisted personnel, the percentage change in force size is 0 percent, 0.1 percent, and 0.6 percent for each Army component, respectively. In the case of the RA, force size is held constant by increasing or reduc-

Table 4.2
Percentage Change in Force Size Relative to the Baseline, by Army Component and COA

COA	RA	USAR	ARNG
Enlisted personnel[a]			
1. RA, USAR, ARNG set to floor	0	0.1	0.6
Officer			
1. RA, USAR, ARNG set to floor	0	7.2	2.1
2. Optimize RA, USAR, ARNG	0	–0.2	–3.0
3. Optimize RA, USAR; ARNG set to floor	0	0.1	–3.1
4. Optimize USAR; RA and ARNG set to floor	NA[b]	NA[b]	NA[b]

[a] COAs 2–3 are not relevant for enlisted personnel, because retention is sustained under COA 1.
[b] COA 4 is infeasible so percentage changes in force size were not estimated.

ing accessions as retention changes under each COA. Thus, the table shows no change in RA force size, by assumption.

The results differ for officers, however, as we also found in our earlier BRS analysis for the Army. The CP multiplier floors are too low for officers. RA officer retention falls and fewer RA officers complete 20 YOS, as shown in the left panel of Figure 4.4—the red line falls below the black line. The purpose of the CP is to draw junior officers to midcareer; once they reach midcareer, the DB retirement plan and the 20-year vesting further draw them to complete at least a 20-year career. Set at the floors, the CP multipliers are an insufficient draw for officers. Instead, more officers choose to participate and complete 20 YOS in the USAR. As shown in the left panel of Figure 4.3, steady participation in the USAR among officers with prior AC service increases between YOS 5 and 20. Overall USAR participation is estimated to increase by 7.2 percent (Table 4.2). On the other hand, participation in the ARNG would increase by less—by 2.1 percent, as shown in the right panel of Figure 4.3 and in Table 4.2. These results show the interrelationship between the CP multipliers across the three Army

Figure 4.4
COA 1: RA Officer and Enlisted Personnel Retention Under the BRS Versus Baseline

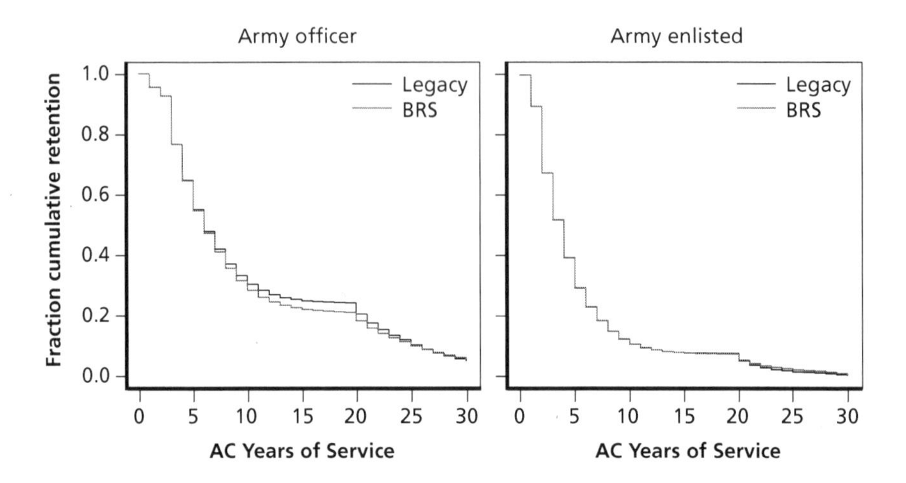

components and their effects on RA retention, and USAR and ARNG participation. They also suggest that the USAR enlisted personnel with prior RA service are more responsive to changes in CP than are ARNG personnel.[3] In summary, the simulations show that the BRS would improve USAR and ARNG participation among officers, but at the cost of substantially lower retention of midcareer RA officers.

A natural question is why the results for officers differ from those of enlisted personnel. The difference is attributable to several factors. First, because the BRS reduces the DB multiplier from 2.5 percent to 2.0 percent, an individual in early career or midcareer, looking ahead to military retirement, will expect a lower DB annuity under the BRS. Officers have higher retention and a higher likelihood of reaching 20 YOS, so they are more likely to experience the reduction in the DB multiplier than their enlisted counterparts. Second, officers earn higher basic pay and can expect a larger DB annuity if they become eligible at

---

[3] Although we do not explore this issue here, the results suggest that USAR enlisted personnel with prior RA service are more responsive than ARNG personnel to changes in special and incentive pay targeted in the midcareer.

20 YOS. Consequently, the reduction in the DB multiplier has a larger effect for officers than for enlisted personnel with a given number of YOS. Finally, we estimate differing parameters for officers and enlisted personnel and, in particular, different personal discount factors. We typically estimate a personnel discount factor for officers of 0.94 and for enlisted personnel of between 0.88 and 0.90.[4] This implies that a dollar one year from now is worth more today for officers—$0.94—than for enlisted personnel—$0.88 to $0.90. Thus, a given reduction in future benefits, such as the reduction from the reduced DB multiplier under the BRS, has more of an effect on today's value for officers than for enlisted personnel. Both the CP and TSP help to offset the negative effects of the reduced DB multiplier, but the optimized CP multiplier must be higher for officers than for enlisted personnel.

We next consider COAs 2 and 3 (see Table 4.2). Because enlisted retention is sustained with the mandated CP multiplier floors, these COAs only apply to officers, and as discussed earlier, COA 4 is not feasible for officers, so we do not show results for COA 4. The key result of both COA 2 and 3 is that higher CP multipliers are required for officers to maintain AC retention and RC participation under the BRS relative to the legacy system. Further, we find little difference in retention and participation between COAs 2 and 3.

Under COA 2, CP multipliers for the RA, USAR, and ARNG are determined as a result of an optimizing routine that seeks to jointly minimize the distance between RA retention and USAR and ARNG participation under the BRS versus under the legacy system. Joint optimization accounts for the interacting effects between the components when setting the CP multipliers. Figures 4.5 and 4.6 show the effects of COA 2 on the RC and AC, respectively.

The simulations show that setting the RA multiplier for CP at 8.28 months of basic pay at 12 YOS, and the USAR and ARNG CP multipliers at 2.25 and 1.08, respectively, would sustain RA force size and USAR participation relative to the baseline, given the other ele-

---

4 See Mattock, Hosek, and Asch, 2012; Asch, Hosek, and Mattock, 2013; and Asch, Hosek, and Mattock, 2014.

Figure 4.5
COA 2: RA Participation of USAR and ARNG Officers with Prior RA Service Under the BRS Versus Baseline

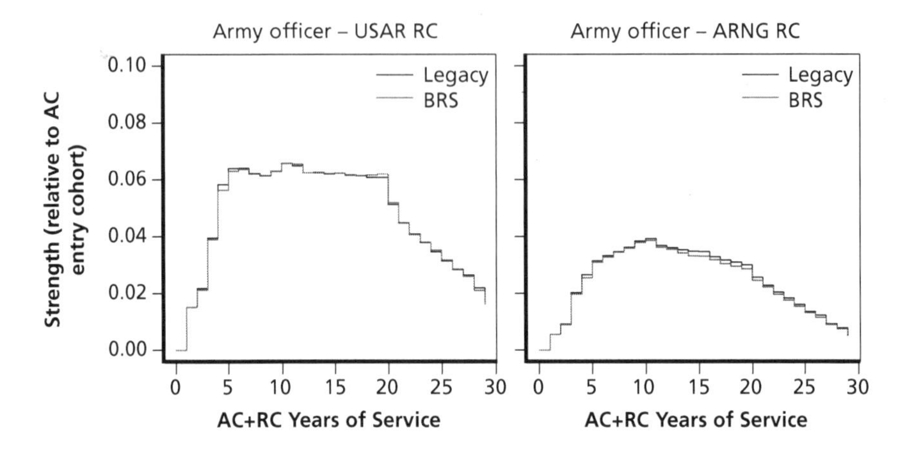

Figure 4.6
COA 2: RA Officer Retention Under the BRS Versus Baseline

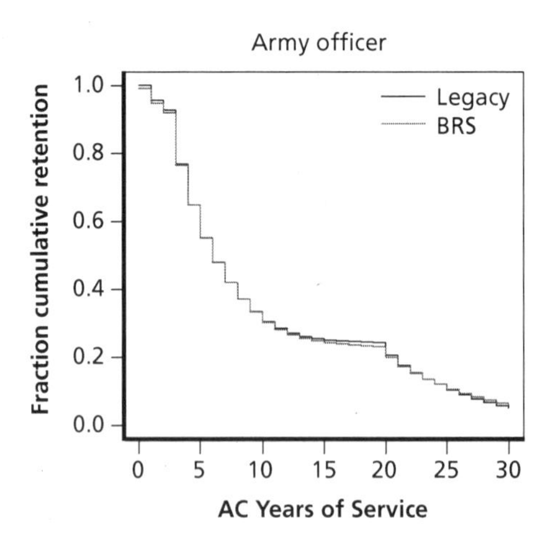

Figure 4.7
COA 3: RA Participation of USAR and ARNG Officers with Prior RA Service Under the BRS Versus Baseline

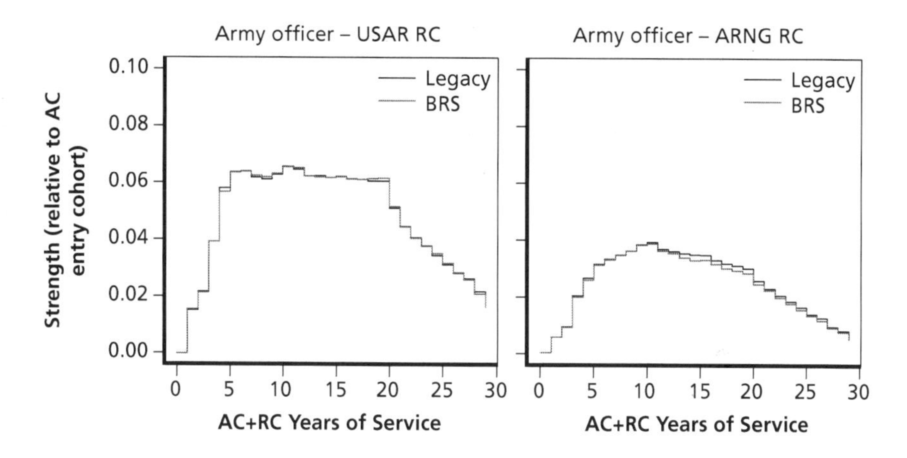

ments of the BRS. ARNG participation would be slightly lower, by 3 percent (Table 4.2).

Figures 4.7 and 4.8 show the results for COA 3 where the CP multipliers are optimized for the RA and USAR, but set to the floor for ARNG. Like COA 2, the results show that multipliers higher than the mandated floor for Army officers substantially reduce the drop in RA retention near 20 YOS, shown in Figure 4.2. That said, holding the CP multiplier for the ARNG at the floor and only optimizing the multipliers for the USAR and RA under COA 3 result in very similar results to the force size changes estimated for COA 2. That is, we find the RA retention (Figure 4.8) and USAR participation (Figure 4.7) are sustained under COA 3 with ARNG participation somewhat lower.

Steady-State Cost

Our simulation capability includes a costing module that enables us to compute BRS-related steady-state costs. Because the DoD actuary computes DB and DC retirement-related costs, the focus of our costing was on CP costs. Table 4.3 shows our estimates of steady-state

Figure 4.8
COA 3: RA Officer Retention Under the BRS Versus Baseline

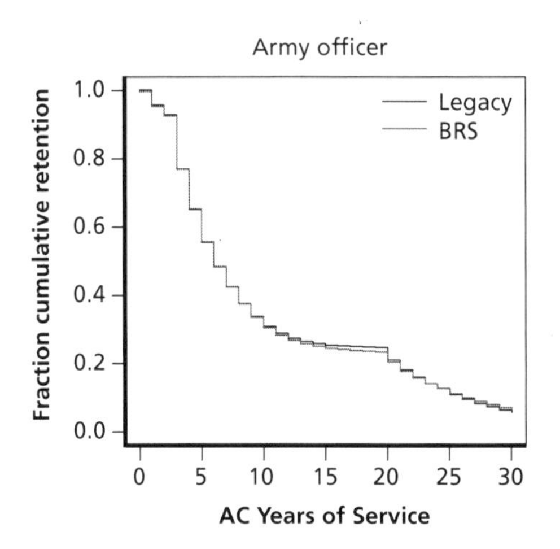

CP costs for enlisted personnel and officers by component: USAR, ARNG, and RA. For enlisted personnel, USAR and ARNG costs are further broken out into those for personnel with and without prior AC service. All dollar figures are in 2017 dollars.

Under COA 1, when CP multipliers were set at the mandated floors, the total cost of CP in the steady state was estimated to be \$104.9 million for RA personnel, \$9.5 million for USAR personnel, and \$18.8 million for ARNG personnel, for a total of \$133.2 million across the three components. It should be noted that because officer retention in midcareer fell short of the baseline for the RA but exceeded the baseline for the USAR and ARNG, the officer cost estimates for COA 1 did not hold retention constant relative to the baseline. This is because the estimated CP costs are affected by both the CP multiplier level, as well as the number of personnel who are at YOS 12 and complete their CP service obligation. Differences in retention will affect the cost estimates.

At the optimized CP multipliers in COA 2 and COA 3, officer RA and USAR force sizes were sustained, with ARNG participation somewhat lower. However, CP costs were also higher for officers;

Table 4.3
Steady-State Continuation Pay Costs, in Millions of 2017 Dollars

	COA 1	COA 2	COA 3
Enlisted total	89.7	89.7	78.9
USAR			
NPS	2.5	2.5	2.5
PS	4.9	4.9	4.9
ARNG			
NPS	10.6	10.6	10.6
PS	7.1	7.1	7.1
RA	64.7	64.7	64.7
Officer total	43.5	149.7	134.6
USAR	2.2	9.3	7.4
ARNG	1.1	2.4	1.1
RA	40.2	138.0	126.1
Enlisted + officer total	133.2	239.4	224.3
USAR	9.5	16.7	14.7
ARNG	18.8	20.0	18.8
RA	104.9	202.7	190.7

NOTE: NPS = non-prior service; PS = prior service.

$149.7 million for COA 2 rather than $43.5 million for COA 1 and $134.6 million for COA 3.

The table also shows the break out of enlisted CP costs for RC personnel with prior versus no prior AC service. The results show that CP costs in the USAR for no-prior-service personnel are lower than for those with prior-service personnel—$2.5 million compared with $4.9 million. This reflects the larger number of USAR members with prior service compared to those with no prior AC service. Furthermore, as shown earlier in Figure 4.1, participation at YOS 12 is relatively low for USAR no-prior-service personnel. The reverse is true for the ARNG where CP costs are higher for personnel with no prior service than with prior AC service. This stems from the larger number of personnel with no prior AC service relative to those with prior service in the ARNG, as well as the relatively higher number of ARNG non-prior service who participate at YOS 12.

We did not compute the costs of the other elements of the BRS, specifically TSP costs and DB retirement costs, because the DoD actuaries estimate them. In addition to the CP costs shown in Table 4.3, the Army will also be required to make TSP contributions on behalf of members, another source of steady-state cost. Offsetting these costs are the savings associated with lower DB payouts (associated with Army retirees) that are accrued by government. Under the BRS, the DB annuity is computed based on a 2.0-percent multiplier instead of the 2.5-percent multiplier under the legacy system. In DoD, retirement costs are funded on an accrual basis, so the DB cost of the BRS to the Army is the accrual charge, computed as a percentage of the basic pay bill. Because of the lower DB multiplier under the BRS, the steady-state accrual charge is lower, thereby producing a cost savings to each of these services. Thus, while the CP cost estimates in Table 4.3 show a source of cost increase associated with the BRS, the lower DB annuity will produce a reduction in costs in the long run. In Chapter Five, we discuss the time pattern of costs in the transition period.

CHAPTER FIVE

Transition Results

The previous chapter showed steady-state results when all members have spent an entire career under the BRS. If the maximum military career length is 40 years, it would take 40 years to reach the steady state. The OCAR was also interested in results for the transition years, especially the first few years, and in particular, in the size of the opt-in population and how the BRS population would grow over time. Opt-in can only occur in 2018, so the size of the opt-in population in the future will depend on the future retention of those who elect the BRS in 2018. For budgeting purposes, the OCAR was also interested in estimates of CP costs over time for the different COAs. To provide these estimates, we extended the DRM simulation capability to simulate the BRS population and CP costs over time for the USAR. We did not pursue developing transition results for the ARNG or RA because the main focus of our analysis is on the USAR. That said, transition results for RA are provided in Asch, Mattock, and Hosek (2017). This chapter summarizes our USAR results. Before presenting the results, we first discuss how and why we extended the DRM.

Extending the Dynamic Retention Model Simulation Capability to Analyze United States Army Reserve Transition

In 2013, we extended the DRM capability to show the effects of compensation policy changes on retention and cost during the transition years for the AC (Asch et al., 2013). In other work, we also considered retention and cost effects for the RC during the transition in analysis of a potential reform to the reserve retirement system to make it more similar to the active system that we conducted for the Army (Mattock, Asch, and Hosek, 2014). Consequently, our 2017 study of the BRS did not consider the opt-in issue or costs for the RC during the transition to the new steady state. We do so in this study, using the modeling for these earlier studies as a foundation.

To assess the effects of the BRS on the USAR, we needed to modify the DRM's mathematical expressions for the value of staying in the military and the value of leaving to explicitly recognize members' years of service or cohort when the BRS policy change was implemented. A member's cohort affects opt-in choice, when CP is paid, and RC participation in future years. Given these expanded expressions, we can derive expressions for cohort-specific probabilities of choosing the RA, the RC, and within the RC, the USAR versus ARNG, and the pure civilian choice in a given period and the evolution of the probabilities over time for each cohort. Furthermore, given an assumed overall force size, we can use these cohort-specific probabilities to compute the retention profile by years of service at a given point in time in the future for those under the legacy system, those under the BRS, and those in the total force. These steps parallel the ones we followed when extending the DRM simulation capability to the transition years for the AC, discussed in detail in Asch et al. (2013).

We had to deal with several complications when developing computer coding that extended the simulation capability of the transition to the USAR. First, BRS enrollment eligibility depends on the number of retirement points, and not on years of service, as with the AC. Second, our simulations of the growth of the BRS population in the USAR over time needed to account for the possibility that a USAR member enrolled in the BRS while the individual was in the RA or ARNG and later joined the USAR. Third, the simulations also needed to recognize that eligible members of the individual ready reserve could enroll into the BRS as a USAR member after 2018 if they are returning to paid status for the first time.

Blended Retirement System Opt-In and Growth in Population over Time

Opt-in behavior among USAR personnel differs for those with and without prior AC service among enlisted personnel, and for officers and enlisted personnel with prior AC service. It also differs by YOS. In this subsection, we show the percentage of USAR personnel that our simulations indicate would elect the BRS in the first year (e.g., in 2018) by years of service and then show simulations of how the BRS population grows for each group, and overall for the USAR.

United States Army Reserve Enlisted Personnel

Figure 5.1 shows the simulated percentage of USAR enlisted personnel that are predicted to elect the BRS by YOS. The orange bars show percentages for personnel with no prior service (NPS) in the RA, and the blue shows percentages for those with prior service (PS). Based on the simulations, we estimate that 100 percent of NPS enlisted personnel with two YOS choose the BRS, while virtually all of the personnel

Figure 5.1
Percentage of USAR Enlisted Personnel Strength That Opt In by YOS at Time of Opt-In Decision

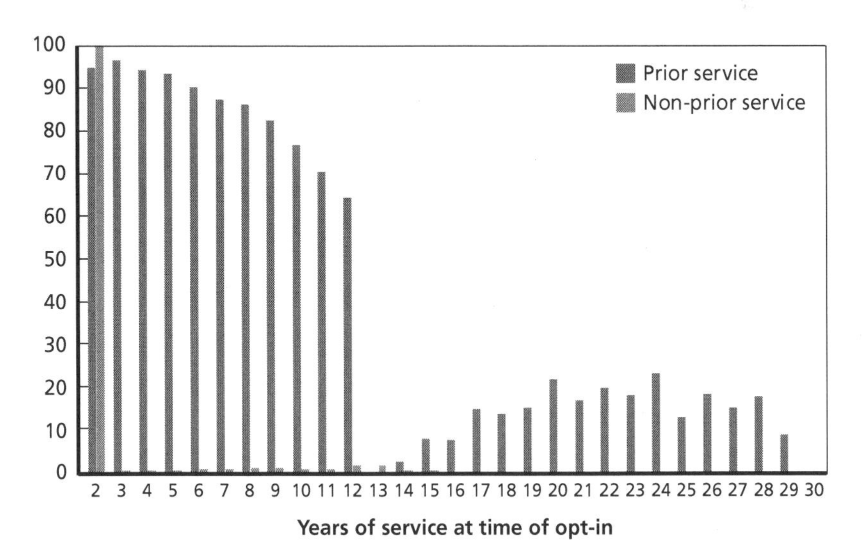

with more than two YOS choose to stay under the legacy retirement system. For PS personnel, opt-in rates are high for those in the initial part of their career, but decline with YOS at the time of opt-in decision until YOS 12. We find positive, albeit lower, opt-in rates for PS personnel with more than 12 YOS.

Opt-in rates are affected by several factors in the model that can operate to increase or decrease the likelihood of opt-in, and the importance of these factors can vary by YOS and between enlisted personnel with and without prior AC service. Those who opt in at later YOS miss out on the TSP contributions that would have been made on their behalf by the agency had they spent an entire career under the BRS, so opt-in rates will tend to be lower for those facing an opt-in decision later in their careers. On the other hand, CP is an inducement to opt in, and it looms larger when individuals are closer to 12 YOS or 4,320 points. Opt-in decisions will be affected by retention over one's career and by the likelihood of reaching key milestones including the beginning of YOS 3, when TSP vesting occurs and when TSP matching begins; YOS 12, when CP is paid; and YOS 20, when DB vesting occurs. Finally, we may observe opt-in even among those with more than 12 YOS. Those with more than 12 YOS miss out on CP but may choose to opt in if they are eligible (have fewer than 4,320 points) to take advantage of TSP contributions made on their behalf.

An NPS USAR-enlisted entrant has a relatively low likelihood of benefiting from the elements of the BRS, given the retention behavior shown in the left panel of Figure 4.1 in Chapter Four. The figure shows that relatively few NPS enlisted USAR entrants are simulated to participate at higher YOS. Only just over half of entrants would reach the start of YOS 3 when TSP matching contributions begin and when TSP vesting occurs, and fewer than 10 percent of entrants would reach YOS 12 when they would receive CP. Finally, only around 5 percent would reach 20 YOS and become vested in the DB under either the BRS or the legacy system. Consequently, only those who are automatically enrolled or have just two YOS are estimated to have elected the BRS.

In contrast, USAR-enlisted entrants with prior AC service have a higher chance of reaching YOS 3, YOS 12, and YOS 20 than do NPS-enlisted entrants, as seen in the right panel of Figure 4.1 in Chapter Four. Figure 5.1 shows that opt-in rates are higher for USAR members with prior AC service for those with fewer than 12 YOS, but opt-in rates decline with YOS and drop sharply after YOS 12.

The OCAR was interested in how the BRS population would grow over time. We show results pertaining to this issue for USAR enlisted personnel in Figure 5.2. The figure shows the simulated percentage of the USAR enlisted force that is under the BRS by year and by whether the force has prior active service. In the first year—2018—we estimate that 37 percent of USAR enlistees with no prior AC service would be under the BRS compared with 56 percent of those with prior AC service. These percentages grow over time for both groups as new entrants are automatically enrolled in the BRS, as those under the legacy system separate, and as those under the BRS are retained. Growth in the fraction of the force that is covered by the BRS is higher in the first seven years for the NPS group but levels off more thereafter, reaching 100 percent in 2057. The growth differences between the PS and NPS groups are due to differences in the retention profiles shown in Figure 4.1 in the previous chapter for each group, and the

Figure 5.2
Percentage of USAR Enlisted Personnel Strength That Is Under the BRS by Year

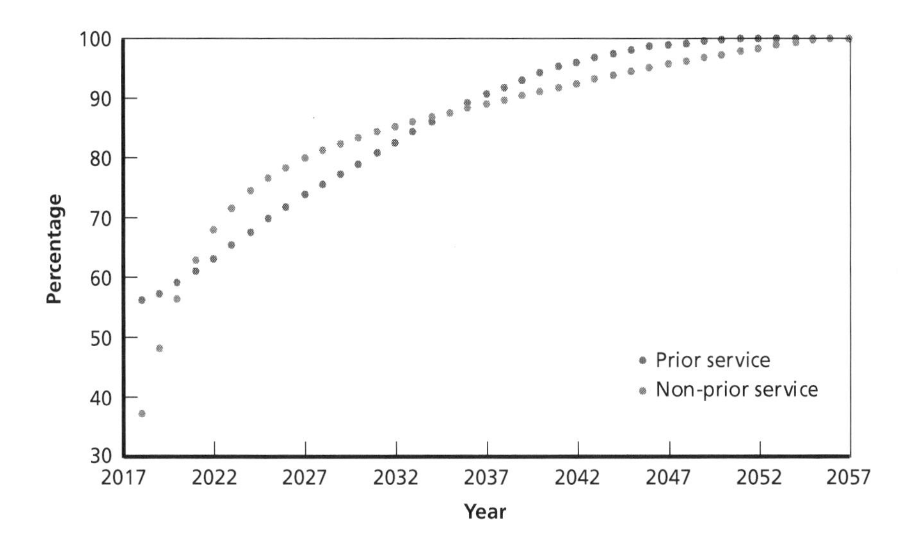

relative size of the population at each YOS within each group. Only the most junior NPS personnel opt in or are automatically enrolled in the BRS, and a large fraction of NPS USAR enlisted personnel have one or two YOS. In contrast, PS enlisted personnel are more evenly dispersed across YOS. The positive effect of the CP multiplier on opt-in behavior is greater under COAs where the CP multipliers are higher.

United States Army Reserve Officers

We also estimate opt-in behavior among USAR officers. Because fewer than 5 percent of USAR officers have no prior AC service, our analysis focuses on those with prior service, and we simulated opt-in behavior among USAR officers under the first three COAs listed in Table 4.1. Under COA 1, the CP multipliers are set to the mandated floor of 2.5 for the RA and 0.5 for the USAR and ARNG. Under COA 2, we optimized the CP multipliers to jointly sustain RA retention and USAR and ARNG participation relative to the baseline. Under COA 3, we optimized the CP multipliers to jointly sustain RA retention and USAR participation and set the CP multiplier for the ARNG to the mandated floor of 0.5.

Figure 5.3 shows the simulated opt-in rates by year of service at the time of BRS implementation, namely 2018 for USAR officers for each COA. The broad pattern is quite similar across COAs as well as similar to the pattern for USAR enlisted personnel with prior AC service (Figure 5.1), though opt-in rates at a given YOS differ across the COAs. Opt-in rates are highest among more junior personnel and decline gradually with YOS until YOS 12, whereupon they drop dramatically thereafter. Opt-in rates will tend to be lower for those facing an opt-in decision later in their careers, because they have missed out on the TSP contributions that would have been made on their behalf by the Army. On the other hand, CP looms larger when individuals are closer to 12 YOS or 4,320 points. The figure shows that opt-in rates are lower for COA 1 than for either COA 2 or COA 3; COA 1 has the lowest CP multipliers while the multipliers are the highest for COA 2.

Figure 5.4 shows the growth of the BRS population over time for USAR officers. In the first year (2018), 15.4 percent of officers are estimated to be under the BRS under COA 1, compared with 29.7 percent

Figure 5.3
Percentage of USAR Officer Strength That Opt In by YOS

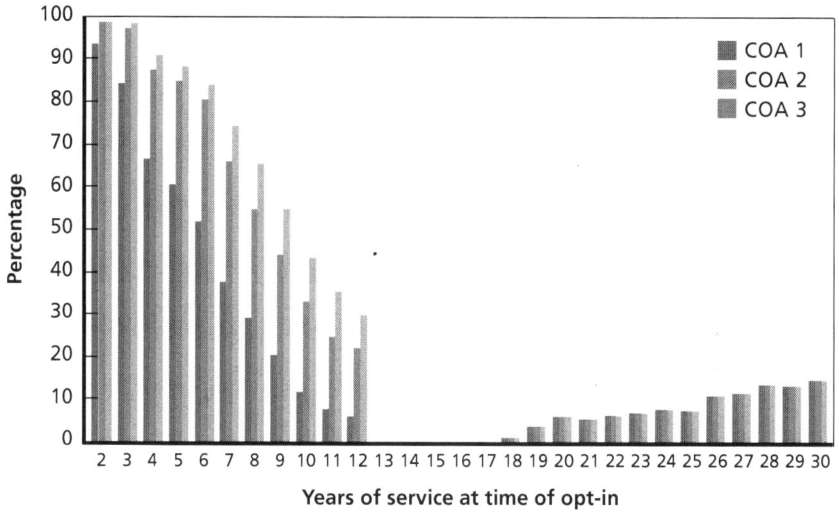

and 25.6 percent for COAs 2 and 3, respectively. The fraction is higher for COA 2 because the CP multipliers are higher than for COA 3 (and COA 1).

Interestingly, between 2018 and 2019, the BRS population declines slightly under COA 1 for USAR officers. This reflects the relative retention (between 2018 and 2019) of the cohorts that elect BRS or are automatically enrolled in BRS in 2018, versus those who are under the legacy system. Those who opt in are predicted to be retained in 2019 at a slightly lower rate than those who stayed under the legacy system. The BRS population under COA 1 increases after 2020. After 2022, the BRS population under COA 1 would increase steadily, somewhat slower than the growth under COAs 2 and 3. Growth is faster under COAs 2 and 3 but the population increases at a decreasing rate with the rate of decrease greater under COA 2. By 2043, or 25 years later, we estimate that just under 70 percent of USAR officers would be under the BRS under COA 1, 86 percent under COA 3, and 90 percent under COA 2.

Figure 5.4
Percentage of USAR Officer Strength That Is Under the BRS, by Year

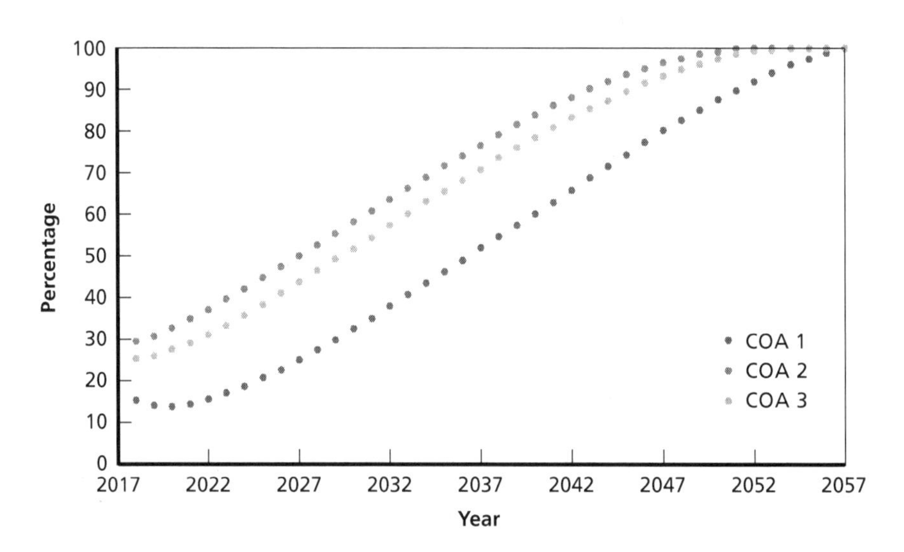

Comparison of Predictions with Actual Opt-In Behavior

DoD released preliminary opt-in rates for USAR and ARNG personnel in January 2019. The preliminary data indicate that USAR opt-in was about 12 percent overall, significantly below the rates predicted with the model, particularly for enlisted personnel. The reason for the lower than predicted opt-in rate is unclear. The DRM assumes members fully understand the elements of the BRS and the financial implications of choosing it over the legacy system. The analysis assumes members choose the system that yields the highest expected value of staying in the USAR, accounting for uncertainty of future outcomes. It is possible members did not fully realize the importance of the opt-in decision or understand the features of the BRS, despite the financial education they received. It is also possible they relied on input from influencers who were older and closer to retirement and who provided input that was not in the best interest of the member. Still another possibility is that the member, although informed about the BRS and the opportunity to opt in, was not required by the service to make a choice. The services differed in their approaches regarding the opt-in decision, and only the Marine Corps required marines to make a choice to either opt in or not. Other services did not require that members make the choice but, in effect, required the member to determine whether to choose. Requiring a choice might have made the decision more salient and induced service members to focus attention on it.

The differences between the predicted and (preliminary) actual opt-in rates provided by DoD suggest that although the DRM fits the data well and predicts retention behavior well, it may require additional information for projecting choices about a new compensation system. In particular it would be useful to have more information on the factors and people that influenced members' choices.

Continuation Pay Costs over Time

Table 4.3 in the previous chapter showed that USAR steady-state CP costs were $9.5 million, $16.7 million, and $14.7 million for COAs 1–3, respectively. These costs include the CP costs for USAR enlisted personnel with both prior and no prior AC service as well as for officers. In this subsection, we show results where we use the DRM simulation capability to estimate the time pattern of CP costs in the transition to the steady state for the three COAs. Figure 5.5 illustrates the results.

Under COA 1, when CP multipliers are set at the mandated floors, we estimate CP costs at the end of the first year of $3 million for the USAR. Costs are estimated to be relatively stable for five years and then grow slowly, reaching $3.7 million by 2027. Costs then increase sharply by $2.4 million to $6.1 million by the end of 2029—corresponding to 12 years after the start of the BRS, when the entering cohort in 2018, who were automatically enrolled in the BRS, reach 12 YOS and receipt of CP. CP costs under COA 1 continue to grow slowly thereafter, eventually reaching the steady-state level of $9.5 million.

Under COA 2, when CP multipliers are jointly optimized for the USAR, ARNG, and RA to sustain retention under the BRS relative to the baseline, we estimate CP costs of $6.7 million in the first year for the USAR. Costs grow steadily until 2027 and then increase by $2.1 million in one year to $11.6 million in 2029 and grow steadily thereafter to the steady state level of $16.7 million. The simulations

Figure 5.5
Total of Enlisted and Officer USAR CP Costs over Time, in Millions of 2017 Dollars

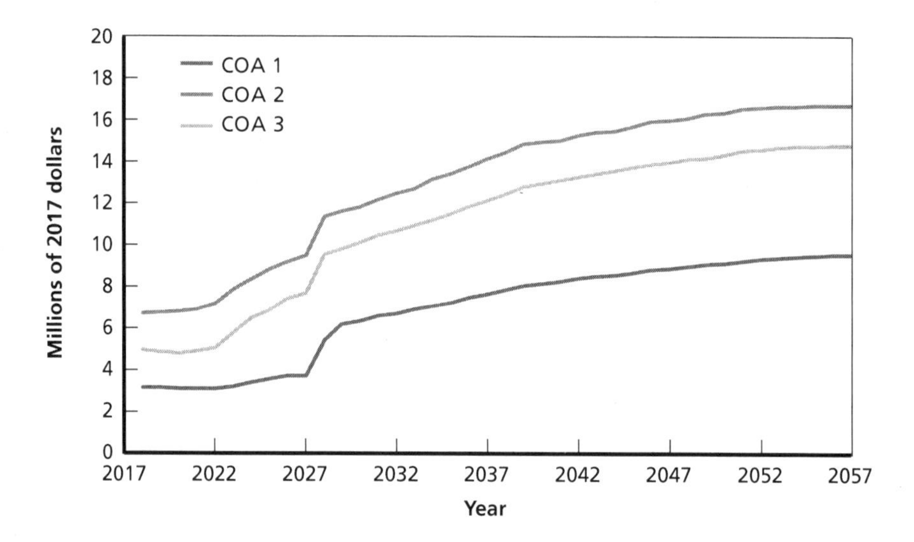

indicate that the time pattern of CP costs under COA 3 is similar to that of COA 2, but costs are a bit lower, reflecting the lower optimized CP multipliers for the USAR and RA (and where the multiplier for the ARNG is set to the floor).

Overall, these results indicate that the time pattern of CP costs is not one of smooth growth. Instead, CP costs increase sharply when entering cohorts that were first automatically enrolled in the BRS in 2018 reach YOS 12 when CP is received. The increase is sharpest under COA 1, the alternative where fewer members choose to opt in to the BRS.

As mentioned earlier, we did not compute TSP costs and DB retirement costs. In addition to the CP costs shown in Figure 5.5, the Army will also be required to make TSP contributions on behalf of members, another source of cost. These increases in costs will be offset, to some extent, by any reductions in the accrual charge associated with the reduced DB benefit under the BRS. How quickly the accrual charge will decline will depend on opt-in behavior across all of the services, since a single accrual charge is set commonly across the services, including the Army. The more members who elect the BRS, the faster the lower DB accrual costs will be realized.

CHAPTER SIX

Concluding Thoughts

Our results show that whether the BRS can support a steady-state force for the USAR and RA that is close to the baseline depends on the CP multipliers that the Army sets. When the CP multiplier is set at the floors of 2.5 and 0.5 for AC and RC personnel, respectively, we find that the baseline force is achievable for enlisted USAR personnel with and without prior active service, given the TSP and DB elements of the BRS. In contrast, for officers, we find that retention is not sustainable for any of the Army components when CP multipliers are set at the floors. In particular, RA officer retention would fall short of the baseline, while USAR participation would increase relative to the baseline; officers who would have completed a career in the AC choose instead to participate in the USAR, assuming they are allowed to do so. To sustain AC officer force size and USAR and ARNG participation CP multipliers must be around eight months' worth of basic pay for the AC, around two months' worth of basic pay for the USAR, and around one month for the ARNG.

Our simulations indicate that opt-in behavior for enlisted personnel depends on whether USAR personnel have prior AC service. Fewer eligible USAR enlisted personnel without prior AC service are predicted to elect the BRS compared with those with prior AC service: 37 percent versus 56 percent. For officers, we estimate that the percentage of the eligible USAR force that elects the BRS increases when the CP multipliers increase. Opt-in behavior in the simulations is affected by several factors including retention and the likelihood of reaching 12 YOS when CP is paid. When CP multipliers are set to the mandated floor, the opt-in rate for USAR officers is estimated to be 15 percent. When CP multipliers are set higher to sustain retention, USAR officer opt-in rate is estimated to be 30 percent under COA 2. We estimate that CP costs in the steady state increase RA costs by $202.7 million per year and increase USAR costs by about $16.7 million per year (both in 2017 dollars), when multipliers are chosen jointly to optimize retention in all three components. We find that the time pattern of CP costs is not smooth—it jumps up after 12 years, when new entrants in 2018 reach eligibility for CP.

We note that the opt-in estimates from the model exceed preliminary estimates of actual opt-in behavior released by DoD in January 2019. For the USAR overall, the opt-in rate among those eligible to elect the BRS was about 12 percent. The reason for the lower than predicted opt-in rate is unclear. The DRM assumes members fully understand the elements of the BRS and the financial implications of choosing it over the legacy system and choose the system yielding the highest expected payoff. As mentioned, perhaps members did not fully realize the importance of the opt-in decision or understand the features of the BRS, or they relied on input from influencers who provided less than satisfactory advice for the member, or they demurred from actually making the decision for or against opting in. The differences between the predicted and (preliminary) actual opt-in rates suggest that the estimated DRM predicts retention behavior well but requires additional information for projecting choices about a new compensation system.

Setting Continuation Pay Multipliers

The BRS includes three components: the TSP, the DB plan, and CP. Under the BRS, the DB multiplier is 2.0, lower than the 2.5 multiplier under the legacy system. The addition of the TSP provides members with the opportunity to vest in military retirement benefits earlier than 20 YOS under the current DB-only legacy plan. The TSP also could help offset any reductions in retention as a result of the lower DB multiplier, though the positive retention effect will be muted by the fact that members are only able to claim their TSP fund after they reach the age of 59 and a half. The purpose of CP is to give the services an additional tool to manage and sustain retention under the BRS. Our simulations show that RA, USAR, and ARNG retention levels are responsive to the CP multiplier levels, and that levels that are higher than the mandated floors required to sustain officer RA retention and USAR participation.

But higher CP levels also increase CP costs, as we show in Chapter Five. These costs are in addition to the cost of TSP contributions made by the Army on behalf of members under the BRS. On the other hand, the DB plan for the Air Force, Army, Marine Corps, and Navy is funded on an accrual basis, and the accrual charge will be decreased in anticipation of lower future outlays for DB. The reduction in the accrual charge would help to offset the increased Army costs associated with CP and the TSP.

Higher CP multiplier levels for officers would also increase the percentage of officers that elect to opt in to the BRS, as was shown in Figure 5.3. Higher opt-in means that future lower costs in the form of lower future DB outlays and lower DB accrual charge will occur sooner. Thus, although CP costs increase when the multipliers are higher, the services realize future cost savings sooner when the multipliers are higher because opt-in rates are higher as well.

An additional consideration is that the RA retention, USAR participation, and CP cost effects of lower CP multipliers for officers only emerge gradually. Figures 4.3 and 4.4 showed that in the steady state, RA retention falls and USAR participation rises when officer CP multipliers are set at the floors under COA 1. But the fall in RA retention, the increase in USAR participation, and the increase in CP costs do not occur immediately. As shown in Figure 5.5 for USAR officers under COA 1, CP costs are relatively constant or grow quite slowly for the first ten years. Only when the 2018 entry cohort reaches 12 YOS do CP costs increase substantially.

These results imply that, if short-term cost considerations are of primary importance, the USAR might set the multiplier at the floor for officers and address USAR participation concerns later when they emerge after ten years. The downside of this strategy is that opt-in rates are estimated to be lower, so future cost savings will occur more slowly.

Furthermore, CP costs for the USAR jump up more sharply after 12 years when CP multipliers for officers are set at the floor. Thus, the increment in future CP costs for the USAR are more significant when CP multipliers are set at the floor. On the other hand, if longer-term savings and sustaining RA retention and USAR participation are of primary importance, then the USAR and RA should set CP multipliers at higher levels for officers.

Wrap-Up

The focus of the analysis summarized in this report was on the effects of the BRS on the USAR. In the process of conducting this analysis, we estimated retention models for enlisted personnel and officers, and developed simulation capability to assess the RA retention, USAR and ARNG participation, and cost effects of alternative compensation policies. The capability considers the interactions between the components and the effects of compensation changes in one component, such as the RA, on the other components, notably the USAR. This capability could be used in the future to assess the effects of other compensation policies of interest, such as retention bonuses in the USAR or accession bonuses for prior-service USAR personnel. It could also be further extended to consider special and incentive pay targeted to specific USAR communities, such as officers in the medical professions, or enlisted personnel in critical occupations. Such analyses could assist the OCAR in ensuring that special and incentive pay and bonuses are in place to sustain USAR force size and justify changes to legislated caps on these pays.

APPENDIX

Dynamic Retention Model Technical Details and Model Estimates

For this study, we estimate dynamic retention models for USAR and ARNG enlisted personnel without prior AC service and an enlisted and officer model of USAR and ARNG personnel with prior AC service. We do not estimate a non-prior-service model for officers because so few USAR officers have no prior AC service. In past studies, we considered RC personnel with prior AC service, but did not distinguish between USAR and ARNG service. Therefore, we must extend our previous work to make this distinction. This appendix describes the extension as well as the model estimates for both prior- and non-prior-service personnel. The dynamic retention model for USAR and ARNG personnel without prior AC service directly corresponds to the non-prior-service model of RC participation discussed in Mattock, Hosek, and Asch (2012).

Extending the Dynamic Retention Model to Include United States Army Reserve and Army National Guard Participation Decisions

In the DRM where USAR members have prior RA service, all members begin their service in the AC. In each time period, the active service member compares the value of staying in the AC with the value of leaving and joining the RC or entering civilian life, and for those choosing the RC, the member chooses participating in the USAR or the ARNG. We use a nested logit approach to capture these decisions, where the active service member is modeled as comparing active service with a civilian/RC nest and the RC participation decision is modeled as comparing civilian opportunities with a USAR/ARNG nest.

Active service has the value

$$V_a + \varepsilon_a$$

where V_a is the nonstochastic portion of the value of the active alternative, and ε_a is the environmental disturbance (shock) term specific to the active alternative, assumed to be extreme-value distributed. We denote the nonstochastic portion of the value of leaving as V_l. The value of leaving involves either joining the RC or entering civilian life, and for those choosing the RC, either joining the USAR or joining the ARNG. Further, these choices are a series of nested choices, so we must consider each nest to develop expressions for the value of leaving.

We denote the USAR choice as u and the ARNG choice as n. The USAR/ARNG nest has the value

$$V_{u/n} + \upsilon_{u/n} = \max\left[V_u + \omega_u, V_n + \omega_n\right] + \upsilon_{u/n}$$

where V_u is the nonstochastic portion of the value of the USAR alternative; V_n is the nonstochastic portion of the value of the ARNG alternative; ω_u and ω_n are the shock terms specific to the USAR and ARNG alternatives, respectively; and $\upsilon_{u/n}$ is the USAR/ARNG nest-specific shock.

Similarly, the civilian/RC nest has the value

$$V_{r/c} + \upsilon_{r/c} = \max\left[V_{u/n} + \upsilon_{u/n}, V_c + \omega_c\right] + \upsilon_{r/c}$$

where the first term is the value of the USAR/ARNG nest; V_c is the nonstochastic portion of the value of the civilian alternative; $\upsilon_{u/n}$ and ω_c are the shock terms specific to the RC and civilian alternatives, respectively; and $\upsilon_{r/c}$ is the civilian/RC nest-specific shock. The common nest shock allows shocks to be common to both the USAR and ARNG, such as unexpected Army policies that affect both components in the same way, while the shocks that affect the USAR and ARNG separately allow for factors that affect each differently, such as unexpectedly good or bad assignments in one component or the other.

The mathematical symbols for nonstochastic values and shock terms are summarized in Table A.1.

Table A.1
Mathematical Symbols for Nonstochastic Values and Shock Terms

Symbol	Interpretation
V_{α}, V_{l}	Nonstochastic values of the AC and of leaving the AC, respectively
V_{u}, V_{n}, and V_{c}	Nonstochastic values of USAR, ARNG, civilian alternatives, respectively
$V_{u/n}$ and $V_{r/c}$	Nonstochastic value of the USAR/ARNG and civilian/RC nests, respectively
ω_{u} and ω_{n}	USAR and ARNG alternative specific shock terms, ω_{u} ~EV[0, η], ω_{n} ~EV[0, η]
ω_{c}	Civilian alternative specific shock term, ω_{c} ~EV[0,λ]
$\nu_{u/n}$	USAR/ARNG nest-specific shock term, $\nu_{u/r}$~EV [0, φ]
$\nu_{r/c}$	Civilian/RC nest-specific shock term, $\nu_{r/c}$~EV[0,τ]
ε_{α}	Active alternative specific shock term, $\varepsilon_{a} \sim \text{EV}\left[0, \sqrt{\lambda^2 + \tau^2}\right]$
ε_{l}	Leaving Active alternative specific shock term, $\varepsilon_{l} \sim \text{EV}\left[0, \sqrt{\lambda^2 + \tau^2}\right]$
ω_{r}	RC alternative specific shock term, $\omega_{r} \sim \text{EV}\left[0, \sqrt{\eta^2 + \varphi^2}\right]$
κ	Standard deviation of active and civilian/RC shock terms
λ	Standard deviation of civilian and RC alternative specific shock terms

The value of staying in the AC is the sum of the individual's taste for active service, γ_a; active military compensation, W_a; and the discounted value of the expected value of the maximum of the AC, civilian, and RC alternatives in the following period. Note that to calculate wages, eligibility for retirement benefits, and so on, we need to keep track of time spent in the AC, time in the RC, and time overall. Thus, we have three time indexes that are each incremented appropriately to reflect the result of the choice in the current period. The time indexes are t_a, t_r, t_t. For example, if an individual serves a year in the AC, then both the time in the AC and the total time will be incremented by one. The nonstochastic part of the value of choosing to stay active is:

$$V_a = \gamma_a + W_a(t_a)$$

$$+\beta E\left[\max\left[\begin{array}{l} V_a(t_a+1, t_r, t_t+1) + \varepsilon_a, \\ \max\left[\begin{array}{l} V_{u/n}(t_a, t_r+1, t_t+1) + \upsilon_{u/n}, \\ V_c(t_a, t_r, t_t+1) + \omega_c \end{array}\right] + \upsilon_{r/c} \end{array}\right]\right]$$

$$= \gamma_a + W_a(t_a) + \beta\, EMaxActive.$$

The mathematical symbols for taste and compensation are summarized in Table A.2.

Table A.2
Mathematical Symbols for Taste and Compensation

Symbol	Interpretation
γ_a, γ_u, and γ_n	Tastes for active service, USAR, and ARNG relative to civilian alternative, $\{\gamma_a, \gamma_u, \gamma_n\} \sim N[M, \Sigma]$
W_a	AC compensation (regular military compensation [RMC])
W_c	Civilian compensation
W_r	RC compensation
β	Discount factor
R	RMC

The value of the USAR/ARNG nest requires expressions for the nonstochastic portion of the value of choosing either the USAR or ARNG. The value of choosing the USAR alternative is the sum of the individual's taste for USAR service, γ_u; RC military compensation, W_r; civilian compensation, W_c; and the discounted value of the expected value of the maximum of the civilian and RC alternatives in the following period where the RC alternative is the USAR/ARNG nest:

$$\begin{aligned} & V_u\left(t_a, t_r, t_t\right) \\ & =\gamma_u+W_c\left(t_t\right)+W_r\left(t_a, t_r\right)+\beta \mathrm{E}\left[\max \left[\begin{array}{l} V_{u/n}\left(t_a, t_r+1, t_t+1\right)+v_{u/n}, \\ V_c\left(t_a, t_r, t_t+1\right)+\omega_c \end{array}\right]\right] \\ & =\gamma_u+W_c\left(t_t\right)+W_r\left(t_a, t_r\right)+\beta \mathrm{E} \mathit{MaxReserveCivilian}. \end{aligned}$$

Similarly, the value of choosing the ARNG alternative is the sum of the individual's taste for ARNG service, γ_n; RC military compensation, W_r; civilian compensation, W_c; and the discounted value of the expected value of the maximum of the civilian and RC alternatives in the following period where the RC alternative is the USAR/ARNG nest:

$$\begin{aligned} & V_n\left(t_a, t_r, t_t\right) \\ & =\gamma_n+W_c\left(t_t\right)+W_r\left(t_a, t_r\right)+\beta \mathrm{E}\left[\max \left[\begin{array}{l} V_{u/n}\left(t_a, t_r+1, t_t+1\right) \\ +v_{u/n}, V_c\left(t_a, t_r, t_t+1\right)+\omega_c \end{array}\right]\right] \\ & =\gamma_n+W_c\left(t_t\right)+W_r\left(t_a, t_r\right)+\beta \mathrm{E} \mathit{MaxReserveCivilian}. \end{aligned}$$

Finally, the value of the civilian alternative is the sum of civilian compensation; any active or RC service retirement benefit that the individual is eligible for, R; and the discounted value of the expected value of the maximum of the civilian and RC alternatives in the following period:

$$
\begin{aligned}
&V_c\left(t_a, t_r, t_t\right) \\
&= W_c\left(t_t\right) + R\left(t_a, t_r, t_t\right) + \beta \mathrm{E}\left[\max\begin{bmatrix} V_{u/n}\left(t_a, t_r+1, t_t+1\right) + v_{u/n}, \\ V_c\left(t_a, t_r, t_t+1\right) + \omega_c \end{bmatrix}\right] \\
&= \gamma_n + W_c\left(t_t\right) + R\left(t_a, t_r, t_t\right) + \beta \mathrm{E} MaxReserveCivilian.
\end{aligned}
$$

To estimate the model, we need to add more structure so that we can derive explicit expressions for V_a, V_u, V_n, and V_c and therefore *EMaxActive* and *EMaxReserveCivilian*. We do so by first assuming that individuals' tastes for AC and USAR and ARNG service are trivariate and normally distributed, with means and standard deviations of taste for AC, USAR, and ARNG service and correlations between AC and USAR service, AC and ARNG service, and USAR and ARNG service. We also assumed that each of the stochastic shock terms in the model, ω_u, ω_n, ω_c, and θ_r, is generated from the extreme-value distribution. When the shocks have the same extreme-value distribution, and in particular have the same variance, then the choice between the nests has a logit form (Train, 2003; Ben-Akiva and Lerman, 1985). From this, we can derive explicit expressions for the distributions of the nest shocks and write explicit expressions for *EMaxActive* and *EMaxReserveCivilian*. We can also derive choice probabilities for each alternative and write an appropriate likelihood equation to estimate the parameters of the model. Thus, we turn to developing explicit expressions for the distribution of the shocks.

The extreme-value distribution $EV[a,b]$ has the form

$$
e^{-e^{\frac{a-x}{b}}}
$$

with mean $a+b\gamma$ and variance $\pi^2 b^2/6$, where γ is Euler's gamma (approximately 0.577), a is the location parameter, and b is the scale parameter. The variance is proportional to the square of the scale parameter, and we use the fact that equal scale parameters imply equal variances. ω_u and ω_n are within-nest errors drawn from an extreme-value distribution $EV[0, \eta]$; and $v_{u/n}$ is the nest-specific error for the USAR/ARNG nest, distributed as $EV[0, \varphi]$. In other words, $v_{u/n}$ can be thought of as a shock that affects both the USAR and ARNG alternatives, whereas ω_u and ω_n affect each alternative separately.

Max $[V_u + \omega_u, V_n + \omega_n]$ also follows an extreme-value distribution with the same scale as for ω_u and ω_n, η, but a different location,

$$\eta \ln\left(e^{V_u/\eta} + e^{V_n/\eta}\right)$$

and mean

$$\eta \ln\left(e^{V_u/\eta} + e^{V_n/\eta}\right) + \gamma\eta.$$

We can write the following:

$$V_{u/n} + v_{u/n} = \max\left[V_u + \omega_u, V_n + \omega_n\right] + v_{u/n} = \eta \ln\left(e^{V_{u/\eta}} + e^{V_{n/\eta}}\right) + \omega' r + v_{u/n}$$

where

$$\omega'_r = \max\left[V_u + \omega_u, V_n + \omega_n\right] - \eta \ln\left(e^{vV_{u/\eta}} + e^{V_{n/\eta}}\right).$$

Define $\omega_r = \omega'_r + v_{u/n}$. It is the sum of two independent, differently distributed extreme-value variables. The error ω'_r is the single error associated with taking the maximum of $V_u + \omega_u$ and $V_n + \omega_n$, and the definition of ω'_r ensures that its mean is $\gamma\,\eta$, corresponding to an extreme-value distribution with location parameter zero and scale parameter η. Further, $v_{u/n}$ is the single error at the nest level. The distributions of ω'_r and $v_{u/n}$ have the same location parameter—zero—but different scale parameters. In general, the variance of the sum of two independent random variables is the sum of the variances, so the variance of $\omega_r = \omega'_r + v_{u/n}$ is $\pi^2(\eta^2 + \varphi^2)/6$ implying a scale parameter for the USAR/ARNG nest of

$$\sqrt{\left(\eta^2+\varphi^2\right)}.$$

It follows that

$$\omega_r \sim EV\left[0, \sqrt{\left(\eta^2+\varphi^2\right)}\right].$$

For brevity, let

$$\lambda=\sqrt{\left(\eta^2+\varphi^2\right)}.$$

Thus, λ is the scale parameter for the shock for the overall USAR/ARNG nest, ω_r, i.e., the RC nest. For parsimony, we also want the shock of the civilian choice, ω_c, to have the same distribution (i.e., the same location and scale parameters), so we set ω_r~$EV[0, \lambda]$.

The civilian/RC nest follows a similar structure. Given our findings for the USAR/ARNG nest, we can write the value of the civilian/RC nest as:

$$\begin{aligned} V_{r/c}+v_{r/c} &= \max\left[V_{u/n}+v_{u/n}, V_c+\omega_c\right]+v_{r/c} \\ &= \max\left[\eta \ln\left(e^{V_{u/n}}+e^{V_{n/\eta}}\right)+\omega_r, V_c+\omega_c\right]+v_{r/c}. \end{aligned}$$

Following the same logic as the USAR/ARNG nest, we get:

$$\begin{aligned} V_{r/c}+v_{r/c} &= \max\left[\eta \ln\left(e^{V_{u/\eta}}+e^{V_{n/\eta}}\right)+\omega_r, V_c+\omega_c\right]+v_{r/c} \\ &= \lambda \ln\left(e^{\eta \ln\left(e^{V_{u/\eta}}+e^{V_{n/\eta}}\right)}/\lambda+e^{V_{c/\lambda}}\right)+\omega'_{r/c}+v_{r/c} \end{aligned}$$

$$\omega'_{r/c} = \max\left[\eta \ln\left(e^{V_{u/\eta}}+e^{V_{n/\eta}}\right)+\omega_r, V_c+\omega_c\right]-\lambda \ln\left(e^{\eta \ln\left(e^{V_{u/\eta}}+e^{V_{n/\eta}}\right)}/\lambda+e^{V_{c/\lambda}}\right).$$

Define $\varepsilon_l = \omega'_{r/c} + v_{r/c}$. The error $\omega'_{r/c}$ is the single error associated with taking the maximum of the USAR/ARNG nest and the civilian alternative. On the other hand, $v_{r/c}$ is the error of the civilian/RC nest. In other words, $v_{r/c}$ can be thought of as a shock that affects both the reserve and the civilian alternatives, whereas ω_r and ω_c affect each alternative separately. Let $v_{r/c}$ be distributed as $EV[0, \tau]$.

The overall shock for the USAR/ARNG and civilian/RC nests is ϵ_l. It is generated by the extreme-value distribution with

$$\varepsilon_l \sim EV\left[0, \sqrt{\left(\lambda^2 + \tau^2\right)}\right].$$

For brevity, define

$$\kappa = \sqrt{\left(\lambda^2 + \tau^2\right)}.$$

Given these distributional assumptions and the implied expressions for the location and scale parameters of the distribution of shocks, we can write the following expressions:

$$EMaxActive = \kappa\left(\gamma + \ln\left[e^{V_{a/\kappa}} + \left[\left(e^{V_{u/\eta}} + e^{V_{n/\eta}}\right)^{\eta/\lambda} + e^{V_{c/\lambda}}\right]^{\lambda/\kappa}\right]\right)$$

$$EMaxReserveCivilian = \kappa \ln\left[\left(e^{V_{u/\eta}} + e^{V_{n/\eta}}\right)^{\eta/\lambda} + e^{V_{c/\lambda}}\right]^{\lambda/\kappa}.$$

Drawing this together, the model can be written as follows:

$$V_a + \varepsilon_a = \gamma_a + W_a + \beta\kappa\left(\gamma + \ln\left[e^{V_{a/\kappa}} + \left[\left(e^{V_{u/\eta}} + e^{V_{n/\eta}}\right)^{\eta/\lambda} + e^{V_{c/\lambda}}\right]^{\lambda/\kappa}\right]\right) + \varepsilon_a$$

$$V_{r/c} + v_{r/c} = \lambda \ln\left(e^{\eta \ln(e^{V_{u/\eta}} + e^{V_{n/\eta}}} / \lambda + e^{V_{c/\lambda}}\right) + \varepsilon_l$$

$$V_{u/n} + v_{u/n} = \eta \ln\left(e^{V_{u/\eta}} + e^{V_{n/\eta}}\right) + \omega_r$$

$$V_c + \omega_c = W_c + W_r + \beta\kappa \ln\left[\left(e^{V_{u/\eta}} + e^{V_{n/\eta}}\right)^{\eta/\lambda} + e^{V_{c/\lambda}}\right]^{\lambda/\kappa} + \omega_c$$

$$\omega_r, \omega_c \sim EV\,[0, \lambda]$$

$$\varepsilon_a, \varepsilon_l \sim EV\,[0, \kappa].$$

Note that the value of staying is given by the first expression while the value of leaving is captured by the three expressions that follow. The last two expressions give the distributional assumptions.

From these expressions, we can derive the transition probabilities. The transition probability is the probability in a given period of choosing a particular alternative, i.e., active, USAR, ARNG or civilian, given one's state. Because we assume that the model is first-order Markov, that the shocks have extreme-value distributions, and that the shocks are uncorrelated from year to year, we can derive closed form expressions for each transition probability. For example, as Train (2009) shows, the probability of choosing to stay active at time *t*, given that the member is already in the AC, is given by the logistic form:

$$\Pr\left(V_a > V_l\right) = \frac{e^{V_{a/\kappa}}}{e^{V_{a/\kappa}} + \left[\left(e^{V_{u/\eta}} + e^{V_{n/\eta}}\right)^{\eta/\lambda} + e^{V_{c/\lambda}}\right]^{\lambda/\kappa}}.$$

We can also obtain expressions for the probability of entering, or staying in, the RC in each subsequent year, given the individual has left the AC.

$$\Pr\left(V_r > V_c\right) = \frac{\left(e^{V_{u/\eta}} + e^{V_{n/\eta}}\right)^{\eta/\lambda}}{\left(e^{V_{u/\eta}} + e^{V_{n/\eta}}\right)^{\eta/\lambda} + e^{V_{c/\lambda}}}.$$

Given this expression, we can also write the probability a member chooses the USAR or chooses the ARNG, given the member has chosen the RC:

$$\Pr\left(V_u > V_n\right) = \frac{V_{u/\eta}}{e^{V_{u/\eta}} + e^{V_{n/\eta}}}$$

$$\Pr\left(V_n > V_u\right) = \frac{V_{n/\eta}}{e^{V_{u/\eta}} + e^{V_{n/\eta}}}.$$

The transition probabilities in different periods are independent and can be multiplied together to obtain the probability of any given individual's career profile observed in the data of active, reserve, and civilian states, and given reserve—USAR or ARNG. Multiplying the career profile probabilities together gives an expression for the sample likelihood that we use to estimate the model parameters using maximum likelihood methods. Optimization is done using the Broyden-Fletcher-Goldfarb-Shanno algorithm, a standard hill-climbing method. We compute standard errors of the estimates using numerical differentiation of the likelihood function and taking the square root of the absolute value of the diagonal of the inverse of the Hessian matrix. To judge goodness of fit, we use parameter estimates to simulate retention profiles for synthetic individuals (characterized by tastes drawn from the taste distribution) who are subject to shocks (drawn from the shock distributions), then aggregate the individual profiles to obtain a force-level retention curve and compare it with the retention curve computed from actual data.

We estimate the following model parameters:

- the mean and standard deviation of tastes for AC, USAR, and ARNG service relative to civilian opportunities (e.g., μ_a, μ_u, μ_n, σ_a, σ_u, σ_n)

- the correlations of tastes for active and USAR, AC and ARNG, and USAR and ARNG, (e.g., ρ_{au}, ρ_{an}, ρ_{un})
- a common scale parameter of the distributions of ω_r and ω_c, λ, a scale parameter of the distribution of ω_u and ω_n, η, and a scale parameter of civilian/RC nest, $v_{r/c}$, τ
- five switching costs, SWITCH1—SWITCH5. These costs are not actually paid by the individual but are implicit in making certain transitions. The first reflects the cost of leaving active duty before the completion of one's initial active duty service obligation. This is an implicit cost to the individual of not fulfilling an obligation of service. The second reflects the cost of transitioning directly from active to reserve service. The third reflects the cost of obtaining a reserve position after being a civilian before one's total service obligation is completed, whereas the fourth reflects the cost of obtaining a reserve position after being a civilian after one's total service obligation is completed. These two may be seen as, in part, representing the difficulty in finding an available reserve position for which the member is qualified in the desired geographic location, particularly when not transitioning directly from active duty. The fifth reflects the cost of leaving the reserve before one's reserve service obligation is fulfilled.

In past implementations of the DRM, we typically estimate a personal discount rate for enlisted personnel and officers. In this study, we fixed the personal discount rates for officers because we found the model fits were better and parameter estimates were more reasonable relative to our expectations based on past research. We set the personal discount factor in this model equal to 0.94 for officers, which is the value we have typically estimated for officers in earlier work. For enlisted personnel we set the personal discount factor to 0.88, in line with our previous empirical estimates for the average population discount factor upon entry into service.

Once we have parameter estimates for a well-fitting model, we can use the logic of the model and the estimated parameters to simulate the AC cumulative probability of retention to each YOS in the steady state for a given policy environment, and specifically the BRS. By *steady state*, we mean when all members have spent their entire careers under the policy environment being considered. The simulation output includes a graph of the AC retention profile, USAR and ARNG by YOS, where YOS for USAR and ARNG counts active service. We show model fit by simulating the steady-state retention profile in the current policy environment and comparing it with the retention profile observed in the data, as shown in Chapter Three.

Estimation and Coefficient Estimates for Enlisted and Officer Personnel

We report the parameter estimates and standard errors separately for USAR and ARNG enlisted personnel with no prior AC service in Tables A.3 and A.4, respectively. Table A.5 shows parameter estimates and standard errors for Army enlisted personnel with prior AC service, and Table A.6 shows the parameter estimates and standard errors for Army officers with prior AC service. The final column in the tables shows the transformed coefficient estimates. As mentioned, to make the numerical optimization easier, we did not estimate most of the parameters directly, but instead estimated the logarithm of the absolute value of each parameter (except for the taste correlation where we estimated the inverse hyperbolic tangent of the parameter). To recover

Table A.3
DRM Estimates for USAR Enlisted Personnel with No Prior AC Service, Beta = 0.88

Parameter	Coefficient Estimate	Standard Error	Z-Statistic	Transformed Value
Lambda	2.432	0.054	44.658	11.376
Mu	2.062	0.027	76.625	–7.861
Sigma	0.578	0.121	4.767	1.783
Switch1	4.245	0.054	78.389	–69.780
Switch2	1.126	0.065	17.464	–3.082

NOTE: Transformed parameters are denominated in thousands of dollars.

Table A.4
DRM Estimates for ARNG Enlisted Personnel with No Prior AC Service, Beta = 0.88

Parameter	Coefficient Estimate	Standard Error	Z-Statistic	Transformed Value
Lambda	6.267	0.551	11.375	526.973
Mu	4.651	0.529	8.791	–104.713
Sigma	5.044	0.559	9.031	155.037
Switch1	8.084	0.551	14.678	–3,243.429
Switch2	5.094	0.552	9.221	–162.993

NOTE: Transformed parameters are denominated in thousands of dollars.

the parameter estimates, we transformed the estimates back to their unlogged values.

All coefficients are estimated in logarithms (multiplying by –1 as appropriate), with the sole exception of ρ. The correlation coefficients in the models of prior-service RC service are measured as the inverse hyperbolic tangent, a convenient way of mapping the real line to a [–1,1] interval. The back-transformed coefficients are reported in the "transformed value" column of the tables; coefficients corresponding to monetary values are measured in thousands of dollars.

We turn first to the models of USAR and ARNG enlisted personnel with no prior AC service: All the coefficients are significant and show the expected sign but differ in magnitude. The parameter estimates for ARNG are much larger than seems credible, while the USAR parameter estimates are much more reasonable. However, this disparity in size may also be due to the wide difference in National Guard programs from state to state, as well as the incidence of contingencies from year to year under which one or a few state governors might call out the National Guard (as in the Los Angeles riots in 1992, or Hurricane Katrina in 2005). The shock term lambda is smaller in the USAR than in the ARNG, possibly reflecting a lower incidence of unexpected year-to-year change. The estimated mean taste for service at entry, mu, is higher in the USAR than in the ARNG, but the larger taste variance

Table A.5
DRM Estimates for USAR Enlisted Personnel with Prior AC Service, Beta = 0.88

Parameter	Coefficient Estimate	Standard Error	Z-Statistic	Transformed Value
Tau	3.211	0.006	508.054	24.811
Lambda	2.583	0.010	251.440	13.238
Eta	−0.097	0.000	−4.851e+06	0.908
Mu1	2.534	0.002	1,089.238	−12.598
Mu2	2.938	0.009	338.721	−18.882
Mu3 delta	0.100	0.005	21.291	−20.877
Sigma1	−0.845	0.140	−6.021	0.430
Sigma2	1.904	0.014	133.868	6.714
Sigma3 delta	0.399	0.008	48.685	10.008
atanhRho12	1.831	0.022	81.685	0.950
atanhRho13	0.311	0.004	78.872	0.301
atanhRho23	0.224	0.004	52.534	0.220
-Switch1	3.590	0.008	432.112	−36.243
-Switch2	2.900	0.016	186.710	−18.174
-Switch3	4.097	0.015	283.144	−60.173
-Switch4	4.187	0.011	384.903	−65.851
-Switch5	3.474	0.025	141.627	−32.260

NOTE: Transformed parameters are denominated in thousands of dollars, with the exception of the Rho (correlation) terms.

sigma for ARNG means that the ARNG population taste distribution completely overlaps the USAR taste distribution—that is, there are individuals in the ARNG who have both higher and lower tastes for service than are found in the USAR taste distribution.

The first switching cost in these no-prior-AC service models corresponds to the cost implicit in reentering RC service from civilian life. This coefficient is negative and significant in both models, and in both models is roughly six times the size of the shock term lambda. The second switching cost in these models corresponds to the cost of leaving before fulfilling one's RC service obligation and is also negative and significant in both models.

Turning next to the model of USAR enlisted personnel with prior AC service, all the coefficients are significant. Tau, the shock term associated with the AC versus RC/civilian nest choice, and lambda, the shock term associated with the civilian versus RC choice, are of the same order of magnitude, about $25,000 and $13,000, respectively, whereas eta (the shock term associated with the USAR versus ARNG choice) is much smaller, about $1,000. This may signal that the choice of USAR versus ARNG participation is less subject to year-to-year disturbance than the choice of whether or not to participate in the RC at all.

Mu1, mu2, and mu3 delta are all significant, and the population mean taste at entry for the USAR, mu2, is only slightly lower than the taste for AC service, mu1, at –$19,000 and –$13,000, respectively. Mu3, the population mean taste at entry for the ARNG, is measured as a "delta" from the taste for the USAR, which allows us to quickly ascertain if there is a significant difference in taste at entry between the two. There is, but the difference is small, about $2,000 between the USAR at $19,000 and the ARNG at $21,000. The corresponding taste variance terms, sigma2 and sigma3 delta, tell a similar story to the one above for the enlisted non–prior-service participation models, in that the variance associated with taste for the USAR is smaller than that for the ARNG, and the relative values mean that the ARNG taste distribution completely overlaps the USAR taste distribution.

Rho12, rho13, and rho23 are the correlations between taste for AC service and USAR participation, AC service and ARNG participation, and USAR and ARNG participation, respectively. The correlation between AC service and USAR participation is high, at 95 percent, whereas the correlation between AC service and ARNG participation is significantly lower, at 30 percent. This may be due to a perceived similarity in some occupations, specialties, and career tracks between the AC and USAR. The correlation between taste for USAR

Table A.6
DRM Estimates for USAR Officers with Prior AC Service, Beta = 0.94

Parameter	Coefficient Estimate	Standard Error	Z-Statistic	Transformed Value
Tau	4.883	0.058	84.746	131.966
Lambda	4.454	0.068	65.893	85.946
Eta	1.744	0.174	10.050	5.723
Mu1	3.251	0.033	97.629	–25.828
Mu2	4.388	0.066	66.404	–80.497
Mu3	0.701	0.056	12.505	–162.245
Sigma1	3.172	0.067	47.052	23.848
Sigma2	3.797	0.082	46.582	44.564
Sigma3 delta	0.716	0.067	10.642	91.148
atanhRho12	1.168	0.100	11.640	0.824
atanhRho13	0.449	0.022	20.700	0.421
atanhRho23	0.489	0.025	19.746	0.454
-Switch1	5.893	0.042	140.013	–362.522
-Switch2	5.302	0.077	69.290	–200.795
-Switch3	5.735	0.091	62.831	–309.513
-Switch4	6.027	0.069	87.159	–414.331
-Switch5	4.789	0.118	40.615	–120.163

NOTE: Transformed parameters are denominated in thousands of dollars, with the exception of the Rho (correlation) and Beta (discount factor) terms.

and ARNG participation is also low, at 22 percent, signaling that individuals with a taste for participation in the RC may have differences in preference for service in the USAR versus ARNG, but there is by no means a negative relationship between preference for one versus the other.

The switching costs, SWITCH1—SWITCH5, are all negative and significant, as expected. They are all roughly the same order of magnitude, ranging from the cost associated with switching from the AC to the RC after having fulfilled one's service obligation, switch2, at –$18,000, to the cost associated with entering the RC from civilian status after having not previously fulfilled one's total service obligation, switch4, at –$66,000. The cost associated with entering the RC from civilian status after having fulfilled one's total service obligation, switch3, is slightly lower, at –$60,000. Remarkably, entering the USAR yields a higher implied switching cost than leaving the AC before having completed one's active duty service obligation, switch1, at –$36,000. This could be a result of fewer positions available for junior USAR personnel who have not completed their active obligation. Leaving the RC before having completed one's reserve duty service obligation is only slightly less costly to the individual, switch5, at –$32,000.

In the model of USAR participation for officers with prior AC service, all coefficients are significant and show a structure remarkably parallel to that seen in the preceding model of enlisted participation. The magnitude of the coefficients is larger, in line with our previous DRM estimates for officers in all services; this is likely due to two factors: The discount factor for officers at entry is larger than that for enlisted personnel, thus the future holds more value, leading to a larger valuation placed on the shock and taste terms; and the fact that both RMC and the civilian opportunity wage are larger for officers.

Tau, the between-nest shock term governing the shock common to both the AC nest and RC/civilian nest, is larger, at $132,000 than the within-nest shock for RC/civilian lambda, at $86,000. Eta, the within-nest shock governing the relative value of USAR versus ARNG, is much smaller, at $6,000, meaning that the choice of participation in either the USAR or ARNG is not typically driven by the shock common to both alternatives.

The estimated mean taste at entry for participation in the USAR, mu2, shows the same relationship to the taste for ARNG, mu3, as was seen in the enlisted model, with mu2 at –$80,000 and mu3 significantly lower at –$162,000. Again, we see that the estimated population variance in taste for participation is significantly greater for ARNG with sigma3 at $91,000 versus USAR (sigma2) at $44,000, and the population taste distributions overlap. The correlation between taste for service in the AC and participation in the USAR is significantly higher than the corresponding correlation between taste for AC service and ARNG participation, with rho12 at 82 percent and rho13 at 42 percent. The correlation in taste for participation in the USAR versus ARNG, rho13, comes in at 45 percent.

The estimated value for switch1, the implied cost of leaving the AC for the RC before having completed one's active duty service obligation, is –$362,000. Entering the RC directly from the AC, switch2, after having completed one's active duty service obligation is significantly less costly, at –$201,000, as one would expect. Entering the RC from civilian status either before (switch3) or after (switch4) having completed one's total service obligation is relatively more costly, at –$310,000 and –$414,000 respectively. Leaving the RC before having completed one's reserve duty service obligation (switch5) is considerably less costly to the individual, at –$120,000.

References

Asch, Beth J., James Hosek, and Michael G. Mattock, *Toward Meaningful Compensation Reform: Research in Support of DoD's Review of Military Compensation*, Santa Monica, Calif.: RAND Corporation, RR-501-OSD, 2014. As of March 1, 2018:
http://www.rand.org/pubs/research_reports/RR501.html

Asch, Beth J., James R. Hosek, Michael G. Mattock, and Christina Panis, *Assessing Compensation Reform: Research in Support of the 10th Quadrennial Review of Military Compensation*, Santa Monica, Calif.: RAND Corporation, MG-764-OSD, 2008. As of March 8, 2018:
http://www.rand.org/pubs/monographs/MG764.html

Asch, Beth J., Michael G. Mattock, and James R. Hosek, *A New Tool for Assessing Workforce Management Policies over Time: Extending the Dynamic Retention Model*, Santa Monica, Calif.: RAND Corporation, RR-113-OSD, 2013. As of November 21, 2016:
http://www.rand.org/pubs/research_reports/RR113.html

———, *Reforming Military Retirement: Analysis in Support of the Military Compensation and Retirement Modernization Commission*, Santa Monica, Calif.: RAND Corporation, RR-1022-MCRMC, 2015. As of March 1, 2018:
http://www.rand.org/pubs/research_reports/RR1022.html

———, Michael G. Mattock, and James Hosek, *The Blended Retirement System: Retention Effects and Continuation Pay Cost Estimates for the Armed Services*, Santa Monica, Calif.: RAND Corporation, RR-1887-OSD/USCG, 2017. As of March 8, 2018:
https://www.rand.org/pubs/research_reports/RR1887.html

Ben-Akiva, Moshe, and Steven Lerman, *Discrete Choice Analysis: Theory and Application to Travel Demand*, Cambridge, Mass.: MIT Press, 1985.

DoD—*See* U.S. Department of Defense.

Mattock, Michael G., Beth J. Asch, and James Hosek, *Making the Reserve Retirement System Similar to the Active System: Retention and Cost Estimates*, Santa Monica, Calif.: RAND Corporation, RR-530-A, 2014. As of May 3, 2018: https://www.rand.org/pubs/research_reports/RR530.html

Mattock, Michael G., James Hosek, and Beth J. Asch, *Reserve Participation and Cost Under a New Approach to Reserve Compensation*, Santa Monica, Calif.: RAND Corporation, MG-1153-OSD, 2012. As of December 10, 2014: http://www.rand.org/pubs/monographs/MG1153.html

Office of the Assistant Secretary of Defense, *Official Guard and Reserve Manpower Strengths and Statistics: FY 2007 Summary*, Washington, D.C., January 2007.

Office of the Under Secretary of Defense, Personnel and Readiness, Directorate of Compensation, "Greenbook," *Selected Military Compensation Tables*, Washington, D.C.: U.S. Department of Defense, 2007.

———, *Population Representation in the Military Services Fiscal Year 2009*, Washington, D.C., 2009.

Train, Kenneth, *Discrete Choice Methods with Simulation*, Cambridge, Mass.: University Press, 2003.

U.S. Department of Defense, "Introduction to the Blended Retirement System," PowerPoint presentation, Washington, D.C., June, 2017. As of January 8, 2019: https://militarypay.defense.gov/BlendedRetirement/